Research Notes in Mathematics

Submission of proposals for consideration

Suggestions for publication, in the form of outlines and representative samples, are invited by the editorial board for assessment. Intending authors should contact either the main editor or another member of the editorial board, citing the relevant AMS subject classifications. Refereeing is by members of the board and other mathematical authorities in the topic concerned, located throughout the world.

Preparation of accepted manuscripts

On acceptance of a proposal, the publisher will supply full instructions for the preparation of manuscripts in a form suitable for direct photo-lithographic reproduction. Specially printed grid sheets are provided and a contribution is offered by the publisher towards the cost of typing.

Illustrations should be prepared by the authors, ready for direct reproduction without further improvement. The use of hand-drawn symbols should be avoided wherever possible, in order to maintain maximum clarity of the text.

The publisher will be pleased to give any guidance necessary during the preparation of a typescript, and will be happy to answer any queries.

Important note

In order to avoid later retyping, intending authors are strongly urged not to begin final preparation of a typescript before receiving the publisher's guidelines and special paper. In this way it is hoped to preserve the uniform appearance of the series.

Titles in this series

1 Improperly posed boundary value problems
A Carasso and A P Stone

2 Lie algebras generated by finite dimensional ideals
I N Stewart

3 Bifurcation problems in nonlinear elasticity
R W Dickey

4 Partial differential equations in the complex domain
D L Colton

5 Quasilinear hyperbolic systems and waves
A Jeffrey

6 Solution of boundary value problems by the method of integral operators
D L Colton

7 Taylor expansions and catastrophes
T Poston and I N Stewart

8 Function theoretic methods in differential equations
R P Gilbert and R J Weinacht

9 Differential topology with a view to applications
D R J Chillingworth

10 Characteristic classes of foliations
H V Pittie

11 Stochastic integration and generalized martingales
A U Kussmaul

12 Zeta-functions: An introduction to algebraic geometry
A D Thomas

13 Explicit *a priori* inequalities with applications to boundary value problems
V G Sigillito

14 Nonlinear diffusion
W E Fitzgibbon III and H F Walker

15 Unsolved problems concerning lattice points
J Hammer

16 Edge-colourings of graphs
S Fiorini and R J Wilson

17 Nonlinear analysis and mechanics: Heriot-Watt Symposium Volume I
R J Knops

18 Actions of fine abelian groups
C Kosniowski

19 Closed graph theorems and webbed spaces
M De Wilde

20 Singular perturbation techniques applied to integro-differential equations
H Grabmüller

21 Retarded functional differential equations: A global point of view
S E A Mohammed

22 Multiparameter spectral theory in Hilbert space
B D Sleeman

24 Mathematical modelling techniques
R Aris

25 Singular points of smooth mappings
C G Gibson

26 Nonlinear evolution equations solvable by the spectral transform
F Calogero

27 Nonlinear analysis and mechanics: Heriot-Watt Symposium Volume II
R J Knops

28 Constructive functional analysis
D S Bridges

29 Elongational flows: Aspects of the behaviour of model elasticoviscous fluids
C J S Petrie

30 Nonlinear analysis and mechanics: Heriot-Watt Symposium Volume III
R J Knops

31 Fractional calculus and integral transforms of generalized functions
A C McBride

32 Complex manifold techniques in theoretical physics
D E Lerner and P D Sommers

33 Hilbert's third problem: scissors congruence
C-H Sah

34 Graph theory and combinatorics
R J Wilson

35 The Tricomi equation with applications to the theory of plane transonic flow
A R Manwell

36 Abstract differential equations
S D Zaidman

37 Advances in twistor theory
L P Hughston and R S Ward

38 Operator theory and functional analysis
I Erdelyi

39 Nonlinear analysis and mechanics: Heriot-Watt Symposium Volume IV
R J Knops

40 Singular systems of differential equations
S L Campbell

41 N-dimensional crystallography
R L E Schwarzenberger

42 Nonlinear partial differential equations in physical problems
D Graffi

43 Shifts and periodicity for right invertible operators
D Przeworska-Rolewicz

44 Rings with chain conditions
A W Chatters and C R Hajarnavis

45 Moduli, deformations and classifications of compact complex manifolds
D Sundararaman

46 Nonlinear problems of analysis in geometry and mechanics
M Atteia, D Bancel and I Gumowski

47 Algorithmic methods in optimal control
W A Gruver and E Sachs

48 Abstract Cauchy problems and functional differential equations
F Kappel and W Schappacher

49 Sequence spaces
W H Ruckle

50 Recent contributions to nonlinear partial differential equations
H Berestycki and H Brezis

51 Subnormal operators
J B Conway

52 Wave propagation in viscoelastic media
F Mainardi

53 Nonlinear partial differential equations and their applications: Collège de France Seminar. Volume I
H Brezis and J L Lions

54 Geometry of Coxeter groups
H Hiller

55 Cusps of Gauss mappings
T Banchoff, T Gaffney and C McCrory

56 An approach to algebraic K-theory
A J Berrick
57 Convex analysis and optimization
J-P Aubin and R B Vintner
58 Convex analysis with applications in the differentiation of convex functions
J R Giles
59 Weak and variational methods for moving boundary problems
C M Elliott and J R Ockendon
60 Nonlinear partial differential equations and their applications: Collège de France Seminar. Volume II
H Brezis and J L Lions
61 Singular systems of differential equations II
S L Campbell
62 Rates of convergence in the central limit theorem
Peter Hall
63 Solution of differential equations by means of one-parameter groups
J M Hill
64 Hankel operators on Hilbert space
S C Power
65 Schrödinger-type operators with continuous spectra
M S P Eastham and H Kalf
66 Recent applications of generalized inverses
S L Campbell
67 Riesz and Fredholm theory in Banach algebra
B A Barnes, G J Murphy, M R F Smyth and T T West
68 Evolution equations and their applications
F Kappel and W Schappacher
69 Generalized solutions of Hamilton-Jacobi equations
P L Lions
70 Nonlinear partial differential equations and their applications: Collège de France Seminar. Volume III
H Brezis and J L Lions
71 Spectral theory and wave operators for the Schrödinger equation
A M Berthier
72 Approximation of Hilbert space operators I
D A Herrero
73 Vector valued Nevanlinna Theory
H J W Ziegler
74 Instability, nonexistence and weighted energy methods in fluid dynamics and related theories
B Straughan
75 Local bifurcation and symmetry
A Vanderbauwhede
76 Clifford analysis
F Brackx, R Delanghe and F Sommen
77 Nonlinear equivalence, reduction of PDEs to ODEs and fast convergent numerical methods
E E Rosinger
78 Free boundary problems, theory and applications. Volume I
A Fasano and M Primicerio
79 Free boundary problems, theory and applications. Volume II
A Fasano and M Primicerio
80 Symplectic geometry
A Crumeyrolle and J Grifone
81 An algorithmic analysis of a communication model with retransmission of flawed messages
D M Lucantoni
82 Geometric games and their applications
W H Ruckle
83 Additive groups of rings
S Feigelstock
84 Nonlinear partial differential equations and their applications: Collège de France Seminar. Volume IV
H Brezis and J L Lions
85 Multiplicative functionals on topological algebras
T Husain
86 Hamilton-Jacobi equations in Hilbert spaces
V Barbu and G Da Prato
87 Harmonic maps with symmetry, harmonic morphisms and deformations of metrics
P Baird
88 Similarity solutions of nonlinear partial differential equations
L Dresner
89 Contributions to nonlinear partial differential equations
C Bardos, A Damlamian, J I Díaz and J Hernández
90 Banach and Hilbert spaces of vector-valued functions
J Burbea and P Masani
91 Control and observation of neutral systems
D Salamon
92 Banach bundles, Banach modules and automorphisms of C*-algebras
M J Dupré and R M Gillette
93 Nonlinear partial differential equations and their applications: Collège de France Seminar. Volume V
H Brezis and J L Lions
94 Computer algebra in applied mathematics: an introduction to MACSYMA
R H Rand
95 Advances in nonlinear waves. Volume I
L Debnath
96 FC-groups
M J Tomkinson
97 Topics in relaxation and ellipsoidal methods
M Akgül
98 Analogue of the group algebra for topological semigroups
H Dzinotyiweyi
99 Stochastic functional differential equations
S E A Mohammed
100 Optimal control of variational inequalities
V Barbu
101 Partial differential equations and dynamical systems
W E Fitzgibbon III
102 Approximation of Hilbert space operators. Volume II
C Apostol, L A Fialkow, D A Herrero and D Voiculescu
103 Nondiscrete induction and iterative processes
V Ptak and F-A Potra
104 Analytic functions – growth aspects
O P Juneja and G P Kapoor
105 Theory of Tikhonov regularization for Fredholm equations of the first kind
C W Groetsch

106 Nonlinear partial differential equations and free boundaries
J I Díaz

107 Variational convergences for functions and operators
H Attouch

108 A layering method for viscous, incompressible L_p flows occupying R^n
A Douglis and E B Fabes

Topics in relaxation and ellipsoidal methods

M Akgül

University of Delaware/Middle East Technical University, Turkey

Topics in relaxation and ellipsoidal methods

π

Pitman Advanced Publishing Program
BOSTON · LONDON · MELBOURNE

PITMAN PUBLISHING LIMITED
128 Long Acre, London WC2E 9AN

PITMAN PUBLISHING INC
1020 Plain Street, Marshfield, Massachusetts 02050

Associated Companies
Pitman Publishing Pty Ltd, Melbourne
Pitman Publishing New Zealand Ltd, Wellington
Copp Clark Pitman, Toronto

First published 1984

AMS Subject Classifications: (main) 90CXX, 05CXX, 15A39
(subsidiary) 90C10, 90C25, 90C30

Library of Congress Cataloging in Publication Data

Akgül, M.
Topics in relaxation and ellipsoidal methods.

(Research notes in mathematics; 97)
Revision of thesis (Ph. D.)—University of Waterloo, Canada, 1981.
Bibliography: p.
Includes index.
1. Linear programming. 2. Convex programming.
3. Relaxation methods (Mathematics) I. Title.
II. Series.
T57.74.A43 1984 519.7′2 84-1746
ISBN 0-273-08634-0

British Library Cataloguing in Publication Data

Akgül, M.
Topics in relaxation and ellipsoidal methods.—
(Research notes in mathematics; 97)
1. Linear programming 2. Convex
programming 3. Algorithms
I. Title II. Series
519.7 T57.74

ISBN 0-273-08634-0

Reproduced and printed by photolithography
in Great Britain by Biddles Ltd, Guildford

Contents

Acknowledgement

This work is based on the author's thesis submitted to Department of Combinatorics and Optimization, Faculty of Mathematics, University of Waterloo, Canada, with the same title in December 1981. I would like to acknowledge the support of a number of institutions and people.

I would like to express my gratitute to NATO System Sciences Graduate Apprenticeship Program, the Scientific and Technical Research Council of Turkey, Middle East Technical University, University of Waterloo and University of Delaware for their partial support during the preparation of this work.

I would like to thank Professors J. Edmonds, A.R. Conn, B. Richmond of University of Waterloo, K. Truemper of University of Texas at Dallas, J.F. Maurras of University of Paris, J.L. Goffin of McGill University, A. Ben-Israel and J. Eastham of University of Delaware for their help and support during the preparation and writing of this work. Members of Waterloo Complexity Group deserve special thanks for sharing the benefits and difficulties of being graduate students at Waterloo. Thanks are due Dr. Feofiloff for various translations from Russian.

Finally I would like to thank to my family: my children, Deniz and Filiz, and my wife Nurcan for the inspiration and sharing all the hardship during this time, and especially to Nurcan, for translating my scrawl into beautiful typescript.

November 1983

1 Introduction

1.1 SUMMARY

This work is concerned mainly with two nonlinear, nonsmooth approaches to the linear programming, and convex programming problems: the relaxation, and the ellipsoidal algorithms. Our interest is on the complexity aspect of these problems and algorithms. We discuss polynomial and pseudo-polynomial algorithms for the exact or approximate solution of these problems.

In the rest of this chapter, we discuss various, non-simplex, non-combinatorial approaches to the linear programming problem, since our approach is a continuation of those ideas. We treat them in the context of classes *P* and *NP*: that is the problems (or languages) which can be recognised by a deterministic or nondeterministic Turing machine.

This work is in two parts. In the first part, which in Chapter 2, we study the relaxation algorithm, subgradient algorithm, and the relationship among them. We give a thorough survey, with emphasis on convergence rates of the algorithms. The basic idea of the relaxation is simple. One ignores some subset of the constraints and tries to solve the new problem; from the new point a new subset is chosen, and the process is continued until a solution is found. The version that we are interested in, which is due to Agmon-Motzkin-Schoenberg [A 3 , M27], takes at each iteration one inequality in a set of linear inequalities and determines the new point on the line segment between the current point and its reflection with respect to the chosen hyperplane. Each rule for the selection of the hyperplane and the point on the line defines a new algorithm. The algorithm has geometric convergence and in special cases is finite. Although the finiteness of the relaxation algorithm for general linear programs with rational coefficients can be guaranteed, it is not polynomial.

We start with a simple extension of the Agmon relaxation method. We move the current point an ε amount inside its projection on the given hyperplane. This will allow us to bound the number of iterations by M^2/ε^2 , where M is a bound on the distance from the initial point to a point inside an ε-perturbation of the polyhedron under study. For totally unimodular linear programs, M can be bounded by a pseudo-polynomial in the size of input. By combining this technique with binary search we obtain a pseudo-polynomial algorithm for a class of linear programs which includes the totally unimodular ones. The above algorithm gives an approximate solution. By going to a basic feasible solution of the underlying polyhedron, one obtains an optimal solution provided that the approximation is close enough. This can be in polynomial time. We give several variants of the algorithm.

In order to obtain a fully polynomial algorithm with totally unimodular matrices, we employ the Edmonds-Karp [E 9] scaling technique. We first scale the original (primal) problem, to a number of primal subproblems. Then we scale the dual of each subproblem. Suppose the dual of a subproblem is solved. With the aid of complementarity a primal solution is determined, and in repeated applications of this process an optimal solution to the original problem is constructed. A modified version (also polynomial) may be applied to any linear program with integral data, and it either produces an optimal integral solution or proves the coefficient matrix is not totally unimodular.

The second part of this work is on the ellipsoidal algorithm. It is based on the previous reports [A 5 -All]. Our efforts are aimed to clarify the nature of the ellipsoidal algorithm in relation to linear and convex programming, and Hachiyan's result. For this, in Chapter 3, we give a rather long discussion on the historical development of the ellipsoidal algorithm, giving credit to Kelley [K 13], Levin [L 11], Mitiagin [M 26], Judin-Nemirovskii [J 7 ,J 8] and Shor [S 11,S 13,S 18,S 19].

Recent surveys by Bland et al [B24], [S5] are highly recommended. Hachiyan's contribution in resolving the question LP ϵ P is discussed in great detail. His characterization of nonsolvability of a linear inequality system, related works of Chernikov [C20,C21], and Eremin [E12], and two independent proofs of the characterization are given. The approximate arithmetic aspect of the algorithm is discussed in detail. In their ingenious, but ad hoc, proof of the Hachiyan's main result, Gács-Lovász [G13] introduced the full dimension hypothesis, and, in lack of it, perturbation. This ad hoc proof is treated as 'the proof'. (See, e.g., [G40], [K19], [C22], [P3]). Furthermore, projection to a lower dimensional subspace is suggested [K9]. We showed that Hachiyan solved the decision problem, not the feasibility problem; and for it one does not need the full dimension hypothesis [A5, A7]. Naturally, Hachiyan does not need the full dimension hypothesis [H4]. Furthermore it is shown that, in the context of general convex minimization, a perturbation of constraints is essential, independent of full-dimensionality of the underlying feasible set. Various proofs of Hachiyan's main result are discussed.

In Chapter 4 several methods of solving linear programs via ellipsoidal algorithm are discussed. We suggest a direct algorithm for linear programming. We introduce a class of matrices characterized by the inverse of their basis. For this class, which contains totally unimodular matrices, we study the complexity of the direct ellipsoidal algorithm. An accelerated (or hybrid) ellipsoidal algorithm is suggested with a polynomial bound. When the current point is feasible, we make a virtual move: using the direction given by the objective function cut, we hit one of the boundaries of the underlying polyhedron, then move to a vertex while improving the objective function. We check optimality or unboundedness at the vertex (assuming non-degeneracy). In the absence of both, the value of the objective function at that vertex is utilized in an objective function cut. The process can be viewed as the generation of a sequence of feasibility problems, with the addition of an inequality due to the objective function. It is demonstrated that this new algorithm is capable of solving some well-known hard linear programming programs almost instantly: namely Klee-Minty [K14], Clausen [C13], Edmonds' Shortest Path [C37], Avis-Chvátal [A21], and Goldfarb-Sit [G39]. A modification of the algorithm for the linear complementarity problem solves hard examples of Murty [M], and Fathi [F5] instantly.

In Chapter 5, the complexity of the ellipsoidal algorithm in obtaining approximate optimal solutions is discussed in various convex programming problems: unconstrained, constrained, linearly constrained, convex quadratic, and linear complementarity problems. Ita applicability in non-convex programming is also discussed.

In Chapter 6 it is shown that one can obtain an approximate optimal solution to a convex-concave function on the product of compact spaces in polynomial time. In particular it yields ε-optimal solution for a convex-concave game over polyhedral sets in polynomial time.

Impact of the ellipsoidal algorithm in polyhedra theory and combinatorial programming is discussed in Chapter 7. In particular we discuss the non-solvability of linear inequality system, minimal representation for a linear inequality system is discussed. A selective review of works of Grötchel-Lovász-Schrijver [G41], Karp-Papadimitrou and Padberg-Rao is given.

Chapter 8 is devoted to recent Yamnitsky-Levin [Y3] algorithm. The algorithm replaces the ellipsoids with simplices. The price for this is the n-fold increase in the complexity. We give a self-contained account of the algorithm. We discuss the selection of best simplex with minimum volume. A comparison with ellipsoidal method is also given.

1.2 CLASSES *P* AND *NP*

With developments in combinatorial programming, automata, and recursive functions, it became necessary to study efficiency of algorithms. On a theoretical basis, a measure of efficiency of algorithms is first suggested by Edmonds in his well-known paper 'Path, Trees, and Flowers' [E 3]. Independently Cobham [C26] studied a similar idea. The term 'computational complexity of algorithms' first appeared to the best of our knowledge, in 1963 [H8], in the context of recursive sequences (see also Rabin [R1]).

For an algorithm, its running time and space used during the course of the algorithm are two different criteria. Running time attracted more attention. We will deal with it. There are two different kinds of running time measure: worst case and average case. We will deal with the worst case complexity. By running time we mean the total number of elementary arithmetic operations needed. We say $f(n) = O(g(n))$, if there exist a constant $c > 0$ such that $f(n) \leq cg(n)$ for $n \geq n_o$, for some n_o.

Since the available digital computers and their programming languages vary considerably in performance, one must model the complexity calculations on a universal model. One such model is a Turing machine (see [A 4 ,G 15]). There are two more basic computer models, namely Random Access Machine (RAM) and Random Access Stored Program (RASP) [A 4 ,G 15,T 2]. Fortunately all these models are polynomially equivalent, i.e. if one algorithm runs on model A with time complexity $f_A(\cdot)$, then there exist a polynomial $g(\cdot)$ such that the same algorithm runs in model B in $f_B(n)$ which is $\leq g(f_A(n))$.

Edmonds called an algorithm <u>good</u> (later term polynomial preferred) if there exists an upper bound on the time complexity of the algorithm which is polynomial in some parameters of the problem. A measure of length of input crystallized in later years. Edmonds [E 3] , also gave an example of such an algorithm, namely the blossom algorithm for maximum cardinality 1-matching. It turned out the complexity bound was a function of the number of vertices, assuming all the numbers used and generated by the algorithm will have the same processing time independent of their lengths. Since it was 1-matching, the right hand-side of the associated LP was fixed for all, so it entered as a coefficient in the complexity function.

This caused some confusion which still continuous, in a lesser degree, even among the some prominent researchers in the field.

Edmonds [E 4 ,E 5 ,E 6] later clarified the term 'size of input' to include a measure of size of largest element and/or its significant digits. Today the convention is to define the size of the input of a problem P as the total space needed to write the problem characteristics in binary encoding, with possible addition of dimension, number of edges etc., as additional parameters. Binary encoding is not restrictive since any k-ary representation, for $k \geq 2$, is a linear function of binary representation. We denote by $\ell(P)$ (ℓ in short) the size of the problem. $\sigma = \sigma(P)$ denotes the largest element in absolute value in the representation of P, e.g. in the case of linear programming σ is the largest element in A,b,c. Clearly $\ell \leq f(\log \sigma)$, where $f(\cdot)$ is a polynomial in $\log \sigma$, n etc., e.g. in the above case $f = (m+1)(n+1)(2+\log \sigma)$.

We say an algorithm for a problem P is of polynomial time complexity (polynomial, in short) if its worst case running time is bounded by $f(\ell)$, for some polynomial in ℓ, n, etc. That is, for all instances of P with the same parameters, $f(\ell)$ is an upper bound for its time complexity. We say that the algorithm is pseudo-polynomial if f is a polynomial in σ rather than $\log \sigma$. We say an algorithm is of exponential complexity if there exists an instance of the problem such that its running time $\geq a^{f(\ell)}$, for some $a > 1$.

Edmonds' Blossom Algorithm [E 3] for (weighted) m-matching is a well-known example of a polynomial algorithm. However the Blossom Algorithm for (weighted) b-matching [P 19], as well as the Balinski-Gomory algorithm [B 4] for the assignment problem, are examples of pseudo-polynomial algorithms. All branch and bound algorithms for general integer programming, as well as the simplex method, are well-known examples of exponential time algorithms.

From the theoretical computer scientist point of view it is quite natural to transform the problems to decision or recognition problems, since they are polynomially equivalent to feasibility and/or optimization problems (see [G15 ,K7]).

Thus we define *P* as the collection of problems for which a polynomial time algorithm exists for their solution. An equivalent definition, in the language of automata theory, is the following: a language L $\in$ *P* if it can be recognized by a deterministic Turing machine in polynomial time.

There is another well-known class in complexity theory: *NP*. For our purposes, we say P $\in$ *NP* if given a description of P, and some extra information, say a candidate solution, we can decide with polynomial effort in the size of the description, i.e. $\ell(P)$ and extra information, whether the candidate solution is in fact a solution of P. Another definition is as follows: a decision problem P $\in$ *NP* if it is solvable by a backtrack search (or branch and bound) algorithm in which the depth of the search tree and the number of computations at each node (subproblem) can be bounded by fixed polynomials in $\ell(P)$, so that for every 'yes' instance there is some sequence of branches that leads to a 'yes' answer in polynomial time [B24]. In the language of automata theory, L $\in$ *NP* if L can be recognized by a nondeterministic Turing machine in polynomial time. Obviously the breadth of a polynomial-depth tree may grow exponentially, such as in the travelling salesman problem. Edmonds [E 5] informally introduced *NP* class through 'good characterization'. In fact if P has a good characterization, then P $\in$ *NP* $\cap$ co-*NP*, where co-*NP* is the complement of *NP*. In other words, P $\in$ co-*NP* if given an instance of P and some extra information, say a candidate solution, we can tell with polynomial effort in $\ell(P)$ that the candidate solution is not a solution of P. In branch and bound context, we can verify that recognition problem has an answer 'no' with a polynomial depth search tree when in fact that is the case. A well-known example of good characterization for linear inequality system is Farkas lemma: given a system of linear inequalities and a dual vector; we can tell with polynomial effort that system has no solution, or we can verify a solution.

We say P is *NP*-complete if (i) P $\in$ *NP*, ii) every other problem Q in *NP* can be transformed to an instance of P, say, P = T(Q), provided this transformation and size of T(Q) is a polynomial function of the size of Q.

Cook, in his important paper [C 29], showed that the satisfiability problem is *NP*-complete. Later Karp [K 7], utilizing this fundamental technique, established a large class of *NP*-complete problems including travelling salesman, graph coloring, set packing, knapsack, set covering, 0-1 integer programming, etc. .

Clearly $P \subseteq NP$. The question of whether $P = NP$ is open; it is often considered as the 'biggest' open problem in theoretical computer science. Clearly LP $\in$ *NP*. Karp [K 7] posed the question whether LP is *NP*-complete. A lot of researchers believed for some time that LP is in a class between *P* and *NP*, and even introduced the concept *LP*-complete (see [D 20]). As is well-known, Hachiyan showed that LP $\in$ *P*.

1.3 NON-SIMPLEX APPROACHES IN LINEAR PROGRAMMING

The simplex method of solution of linear programming is probably the best known and most used technique in Operations Research, if not in all of applied mathematics.

Based on the Gauss-Jordan pivoting technique, it moves from one vertex to another along the edge of the underlying polyhedron, while keeping the monotonicity of the objective function. It is worth noting that Dantzig's thinking is highly influenced by his work on the Neyman-Pearson lemma in statistics (see Dantzig [D 6],Chapter 2.3 and [D10] for the historical development of linear programming). Dantzig reports that the first known solved linear program is given by Fourier in 1826, as an ℓ_∞ solution of a linear inequality system, found by essentially the simplex method. Dantzig also reports that Kantorovich's early work, though not appreciated at its time, includes an incomplete description of the simplex algorithm; and highly advanced discussions of the transhipment problem. Van de Panne and Rahnama [V1], recently announced that the Kantorovich's method is equivalent to a parametric method but that it is also equivalent with the simplex method with a special pivot rule. We would like to point out a 1940 paper by Kantorovich [K 5], which is not cited by Dantzig (as a matter of fact, by no one in western literature), in which he repeats some of the ideas of the simplex method and suggests a Lagrangian type approach in a more general setting. He includes some applications of his method to the solution of continuous linear programs, i.e. linear functionals are integrals, and variables are functions on compact sets.

The simplex method is essentially a combinatorial method and part of combinatorial linear algebra over an ordered field (see Tucker [T11]). The combinatorial theory of linear programming is extended by, among others, Minty [M 25], Rockafellar [R 8], Bland [B 22], and Fukuda [F 11]. This combinatorial structure is one of the strongest as well as the weakest part of the simplex method. Strength is obvious, e.g. the beautiful theory of oriented matroid programming. The weakness comes from the local character of the simplex method; it is myopic. Klee-Minty [K14], Jeroslow [J 2], Clausen (see [C13]), Avis-Chvátal [A 21], Edmonds (see [C 37]), Zadeh [Z 1], and Goldfarb-Sit [G 39], (see also Blair [B 21]) exhibited, for various pivoting rules of general linear programs or network flow problems, the exponential behaviour of the simplex method. However the simplex method is finite when necessary precautions are taken to prevent cycling, or even without it (see Bland's rule [B 23]).

On the other hand there have always been non-finite, mostly nonlinear type algorithms for linear programming. Interestingly one such method, the ellipsoidal algorithm, turned out to be the first polynomial algorithm for linear programming.

However one should point out that there is a new pivoting rule for which neither exponentiality nor polynomiality is proven: choose the incoming variable which has been in the basis previously the least number of times (Zadeh [Z 2]).

In 1951 Brown and Koopmans [B31] suggested a feasible direction, non-vertex following algorithm for LP. Next to it Brown [B32], suggested the fictitious play for the zero-sum game. It is well-known that an LP can be transformed polynomially to a symmetric zero-sum game. The basic idea is an averaging process: use the opponents past history to reduce the pay-off matrix to a vector and then choose a pure strategy. It was conjectured that in the limit the average of the pure strategies will give the optimum strategies. This is rigorously proved by Robinson in [R 6]. Danskin [D 1], extended this approach to the solution of continuous games over the product of compact spaces.

In 1950 Brown and von Neumann [B 30], suggested a differential equation approach to solve (symmetric) finite zero-sum games. They introduced a differential equation for the strategy vector using an independent variable t. They proved that as $t \to \infty$, any limit point of the strategy vector must be an optimal strategy for the game.

In 1954 von Neumann [V 5] published another iterative method to solve a zero-sum game. Suppose A is an $m \times n$ matrix. One carries out calculations in $(m+n+1)$ dimensions, the last being the value of the game. The basic idea is to reduce ϕ, the sum of the squares of infeasibilities associated with the linearly inequality system. At each iteration, by normalizing the current infeasibilities one obtains another pair of strategy vectors. By taking a convex combination of these two strategy pairs suitably, to minimize ϕ, one obtains the next $(m+n+1)$ vector. von Neumann showed that this method converges to an optimal solution. Moreover he gave the following estimate: if σ is the largest element in A, then $\phi^k \leq \frac{(m+n)4\sigma^2}{k}$. This implies that in order to obtain an

ε-approximate solution to the corresponding linear inequality system, $\varepsilon \leq \frac{2\sqrt{m+n}\ \sigma}{\sqrt{k}}$ must be satisfied. Recently Hachiyan [H 1], showed that this bound can not be improved.

Various algorithms for minimax problems can be used to solve LP problems, but this approach is not followed; probably it would not be efficient. Danskin [D 4] , recently suggested an algorithm for a linear inequality system which is based on game theory concepts: directions are found through a game.

Sadovskii [S1, S2] presented a game theoretical solution for convex programming, in particular linear programming problems. He works with the Lagrangian function, and tries to find a saddle point of the Lagrangian. It is a primal-dual type algorithm. One unique feature of the method is the lack of line search. New iterates are obtained by solving relatively small size matrix games.

Gol'shtein [G44] applied a modified Lagrangian approach developed by Gol'shtein and Tretyakov to linear programming problem. Since it is an iterative method, it is best suited to find an approximate solution of the problem. He reports that he managed to improve the rate of convergence of the algorithm, and his method becomes more effective than simplex method for medium to large size problems based on limited amount of experiments. Malkov, Padchin and Sokolov [M 4] reports results of experiments for a similar algorithm. See also Belov [B 12] and Poljak and Tretyakov [P16].

The simplex method can be seen as a feasible direction algorithm in the mathematical programming context. Indeed the projection method of nonlinear programming, when specialized to LP, is the simplex method [R11]. Recently, Conn [C 28], applied his non-differentiable exact penalty technique ([C 27]) to LP. For a fixed penalty parameter, the algorithm minimizes the unconstrained penalty function. The algorithm moves from one not necessarily feasible, basic point to another basic point of the underlying linear inequality system. When the minimum is not feasible, the penalty parameter is adjusted.

The method is finite, since the number of basic points of the linear inequality system is finite. It is known that there is a treshold value of the penalty parameter, beyond which the minimum point of the penalty function coincides with that of LP. Bartels [B 9], refined the above method in a terminology suitable to the simplex method. See [V3,V4,K29,B12]for computation of penalty vectors.

Reduced gradient method of nonlinear programming has been applied to linear programming by several resourchers: Lemke [Z12], Wolfe [W5]. Recently Kallio and Porteur [K 2] introduced a class of algorithms based on reduced gradient, which are also pivotal and finite. Kallio and Orchard-Hays [K 3] reported some experimental results with reduced gradient method.

Cortey [C 31] introduced an algorithm for linear inequality system by solving sequence of (unconstrained) least-square solution of linear equality system. In an m×n system he introduces 2m new variables, and transform the linear inequality system into 2m ×(2m+n) equality system which is nonlinear. By fixing half of the variables alternatively, he aims to minimizes square of residuals. Even though, he claims polynomiality of the algorithm: there is no proof of it, its an iterative system.

Of course there are numerous relaxation and subgradient methods, most of which will be reviewed in Chapter 2. The relaxation method started with works of Agmon [A 3], and Motzkin and Schoenberg [M 27]. In the Soviet Union Eremin, Bregman, Bulavskii, etc., suggested several relaxation or similar methods for LP. Oettli [O1-2],Kumar [K27], and many others suggested relaxation methods for the solution of LP.

Maurras in 1978 suggested to solve LP via a minimax, or minimization of a convex non-differentiable function [M12]. He suggested two ways to solve the minimization problem. One way is an accelerated relaxation method [M13,J 1], the other one is Wolfe's conjugate subgradient method [W 6]. He showed that for a class containing totally unimodular LP's, the method is pseudo-polynomial. His method has been recently developed to a polynomial algorithm by Maurras et al [M14], which we will describe in Chapter 2. We should also mention that Bland and Edmonds [B25] has a combinatorial algorithm for totally unimodular LP's which is also polynomial.

Recently Mangasarian introduced two nonlinear methods for LP. In [M 8] he perturbs the linear program to a quadratic program (QP), by adding the term $\varepsilon||x||^2$, for small enough ε, then solves it with relaxation methods. In [M 9] he first looks at the dual of minimization of sum of the norms of primal, dual, and slack variables. He then transforms it to an unconstrained, convex, piecewise quadratic, globally differentiable function. He solves it with gradient methods or with an (over) relaxation algorithm. See also [Z13].

There is a particular area for which the relaxation methods work very well: large scale linear programming. What constitutes a large scale LP seems to change depending on availability of computing facilities (see discussion in Shor [S19], Charnes et al [C13]). An appealing feature of the relaxation method is that only the original data and the current iterate need to be stored, and any accumulated roundoff error is automatically compensated for. It is also possible to avoid function evaluations at each iteration, because availability of a covariance matrix (that is pairwise innerproducts of defining normals) provides an easy update formula. Probably the best known example of successful application of the relaxation method is the Held-Karp application in travelling salesman problem [H 9]. Similar successes are reported in Held-Wolfe-Crowder [H11], Crowder [C 35], and Goffin [G 26 ,G 31, G 34]. Minoux [M 24], used a subgradient method to solve large network synthesis problems with multicommodity flow requirements. One general method for large scale problems is the Dantzig-Wolfe decomposition [D 7 ,L 1].

Ben-Israel and Charnes [B13] introduced an algorithm for a class of interval linear programs using 1-inverse (see Ben-Israel and Greville [B 15] for generalized inverse). This algorithm lead to suboptimization method of Robers and Ben-Israel [R 5]. Later Charnes et al [C12] gave a primal algorithm for interval linear programming. Method is a finite, pivotal one. It is competitive with simplex method and for some class of problems it is better than simplex method, see [G53]. However, theoretically interval linear programming method of [C2 ,R5] is shown to be equivalent to simplex: in the sense that one can generate same sequence of vertices on a given LP. They differ on selection of incoming variables, and handling of degeneracy.

Makarenkov [M 3] introduced a new pivotal, finite, non-extremal algorithm, which even allows interior points to be generated during the algorithm. It is a generalization of simplex method, and uses 1-inverses (for an exposition, see Akgül [A13]).

Recently Cooper and Kennington [C30] and Sherali, Soyster and Bainer [S 7] studied non-extreme point/nonadjacent extreme point method for linear programming. Sherali et al [S 7] concludes that based on their experiments on random problems simplex method has a better performance.

Recently, there is a renewed interest in pivotal methods in Soviet Research, [B 1], [T10], [A1], [G1 - G9], [F1 - 4], [M 3]. Previously, there has been some attempts to change 2 or more variables simultaneously from the basis, see Paranjape [P8], and [C30]. Tuniev [T10], introduced a generalization of simplex method allowing several non-basic variables to increase simultaneously. The algorithm uses a generalized Jordan elimination procedure. Acil'diev [A 1] also introduced a similar method allowing several non-basic varia variables to increase simultaneously. Tuniev and Acil'diev algorithms are similar to Makarenkov algorithm, but each has different algebraic machinery.

Gabasov and Krillova, and coworkers, and Fam introduced several algorithms for linear programming. In a three volume text [G3 -5], Gabasov and Krillova introduced primal, dual and primal-dual algorithms for linear programming. They are suboptimal methods, and estimate of non-optimality is easily calculated. Their basic concept is the supporting plan (solution) corresponding to a linearly independent columns as opposed to basic feasible solution of the simplex method. It does not allow cycling. The fact that they are pivotal methods guarantees finiteness: given $\varepsilon > 0$, these algorithms find $\bar{x}$. Such that $\bar{x}$ is feasible and $cx^{+} - c\bar{x} \leq \varepsilon$ for a maximization problem where x^{+} is a solution of the problem. Several extensions are given by Gabasov et al [G 1], and Fam [Fl - 4]. Gabasov, Krillova and Kostyukova [G 12] has managed to extend the main idea of these algorithms to a linear optimal-control problem.

There is class of problems for which there is a (genuine) polynomial time algorithm exists. Each inequality contains at most 2 variables: see Megiddo [M 21] and its references. Megiddo has some novel ideas for linear programming; [M20-23]. He has a linear algorithm in terms of number of constraints, when the number of variables is fixed.

There has been some important development for the simplex method for some special problems. Hung [H22] developed a polynomial primal simplex algorithm for assignment problem (AP) using strongly feasible basis (see Cunningham [C37] or Barr et al [B 8]). Balinski [B 5] gave a polynomial dual simplex method which is better than Hungarian method for AP. Ikura and Nemhauser [I 1] gave a primal simplex for maximum weighted vertex packing on bipartite graphs. For unit costs, algorithm is polynomial in n, number of nodes in the graph. For general case they apply scaling to the cost vector, [R10].

Expected behaviour of simplex method is also settled for some versions. Dantzig [D9] proved polynomiality of expected number of simplex pivot with a convex constraint. Smale [S 35] proved for Lemke's method when LP is casted as a linear complementarity problem. Borgwardt [B26-27] proved for his Schattenecken algorithm. (For recent developments see program of ORSA/TIMS Orlando, 1983 meeting.)

Conn's [C28] method can be seen also minimization of polyhedral convex function once the penalty parameter is fixed. Recently, Gunn [Z11], presented an unconstrained dual to an LP via Fenchel duality, then solved the latter as a piecewise linear optimization. Fourer's [F8] 'semi-linear' programming is another piecewise linear optimization algorithm which covers the usual linear programming. Complementary pivoting algorithm of Lemke [L10] and others are also piecewise linear optimization algorithm. Recently Gould [Z10] presented some experimental result with this approach and usual simplex method: they are competitive.

Reference System

Equations and theorems are numbered within each section. Theorem 1 or (1) refers to those items in the same section. Theorem 3.5 or (3.5) refers to Theorem 5 or equation (5) in section 3 of the same chapter, whereas theorem 2.3.5 refers to theorem 5 of section 3 in chapter 2.

For Russian journals which are systematically translated into english, such as Cybernetics, Doklady, etc. only the english source is given in the references. For others both sources are given when available.

2 Polynomial algorithms for a class of linear problems

In this chapter we study the relaxation algorithm and subgradient algorithm. The relationship among them and the convergence rate, with special emphasis on the Soviet research, is given. The algorithms have geometric convergence rate, and sometimes finite convergence is attained. Although the relaxation algorithm for the decision problem of the linear inequality system with rational coefficients can be shown to be finite, it is not polynomial in general.

We review the basic properties of totally unimodular matrices, since they will be needed later in the chapter. When a feasible, non-basic point to a linear program is known, one is interested finding a basic feasible point of the underlying polyhedron with a better objective function. We give a simple polynomial algorithm, which is simplex based. The algorithm will be needed later in the chapter and in Chapters 3 and 4.

In the main body of the chapter we first give various pseudo-polynomial algorithms for a class of linear programs. They are based on a modification of the relaxation method: over relaxation with a positive parameter. We give two different proofs of the basic lemma, hopefully in order to gain more insight. We also discuss algorithms for equality constraints. In the rest of the chapter a fully polynomial algorithm is presented. It employs the Edmonds-Karp scaling technique extensively. We first scale the primal problem to a number of subproblems. Then we scale the dual of each subproblem. We end up with a set of problems whose right hand side is a unit vector, and coefficients in the objective function are bounded by 4, independent of the dimension of the problem. Solving as many such subproblems as the length of the right hand side times the length of the objective function of the original problem guarantees polynomiality.

2.1 RELAXATION AND SUBGRADIENT METHODS

Relaxation methods have been used for solving systems of linear equations since the early forties (see Southwell [S37], Allen [A16]). The method can be traced back to Kacmarz, etc.

The method was extended for systems of linear inequalities in 1954 by Agmon [A 3], Motzkin and Schoenberg [M 27], developing the ideas of Motzkin. The connection between equalities and inequalities was explicitly exhibited in [A 3].

Consider a system of inequalities:

$$\ell_i(x) = a_i x - b_i \leq 0 , \quad i \in I . \tag{1}$$

For any x, which is not a solution, choose a violated constraint, i.e. i such that $\ell_i(x) > 0$. Then a typical iteration of a relaxation algorithm will be

$$x^+ = x - \lambda \ell_i(x) \frac{a_i}{||a_i||^2} , \quad \lambda \in (0,2] . \tag{2}$$

There are a family of algorithms depending on λ and the way index i is choosen. Let $f(x) = \max_i \ell_i(x)$, and $f_1(x) = \max_i \ell_i(x)/||a_i||$. Clearly when $||a_i|| = 1$, we have $f_1(x) = f(x)$. Obviously x is a solution of (1) iff $f(x) \leq 0$, or $f_1(x) \leq 0$. For $\lambda = 1$, the method is called orthogonal projection when index i is choosen such that $f_1(x) = \ell_i(x)/||a_i||$; that is $\ell_i(x)$ represents the furthest hyperplane from x. If i is choosen such that $f(x) = \ell_i(x)$, then the method is called the maximal residual method. For $\lambda < 1$ (respectively $1 < \lambda \leq 2$), and $f_1(x) = \ell_i(x)$, the method is called under (over) projection or relaxation method. If the index is choosen systematically by:

$$i(r + 1) \equiv i(r) + 1 \pmod n ,$$

then it is called cyclic or systematic projection. When $\lambda = 2$, with $f_1(x)$, it is called the reflection method. In general one can choose any violated inequality, and any $\lambda \in (0,2)$. All the arguments stay valid.

From now on, unless otherwise stated we will assume that $||a_i|| = 1, \forall i$.

<u>Lemma 1:</u> (Agmon) Let $x,y \in R^n$ such that $\ell_i(x) > 0$, $\ell_i(y) \le 0$, $x' = x - \ell_i(x) a_i$, and $x^+ = x + \lambda(x'-x)$. Then:

$$||x^+ - y||^2 \le ||x-y||^2 - |1-(1-\lambda^2)| \; ||x'-x||^2 . \qquad \square$$

Since $P = \{x: Ax \le b\} = \{x: \ell_i(x) \le 0, \forall i\}$, the above lemma shows that x^+ is closer to P than x. A sequence satisfying the lemma is called a Fejer sequence. In 1922 Fejer showed that the set of closest points to set A is identical to the convex hull of A. Thus the basic observation is that the sequence generated by a relaxation algorithm is Fejer-monotone with respect to the solution set of the linear inequality system.

Agmon proved the following:

<u>Theorem 2:</u> Let $P \ne \phi$. For $\lambda \in (0,2)$, the sequence x^k defined by the relaxation algorithm converges to $x \in P$. Furthermore, $\exists \theta \in (0,1)$ such that

$$||x^k - x|| \le 2d(x_o, P)\theta^k , \quad \forall k. \qquad \square$$

Agmon, for $\lambda = 1$, defined $\theta = \sqrt{1-\mu^2}$, and gave a complicated formula for μ. Recently Todd [T 5], obtained a simpler formula for $1/\mu$ based on quadratic programming duality.

Motzkin and Schoenberg [M 27], mainly studied the case for $\lambda = 2$. They proved:

<u>Theorem 3:</u> Let $P \ne \phi$, and let P be full-dimensional. Let x^k be a sequence obtained through a relaxation algorithm. Then

(i) If $\lambda\epsilon(0,2)$ then either x^k terminates, or else x^k converges to a boundary point of P.

(ii) If $\lambda = 2$ then the sequence x^k always terminates.

Here to terminate means for some finite k, $x^k \epsilon P$. Let L_r be a given flat of dimension r in R^n, and let $p \notin L_r$. Let X be the set of points x such that

$$||x-a|| = ||p-a|| , \quad \forall a \epsilon L_r .$$

Then X is a spherical surface S_{n-r-1}, dimension n-r-1 having L_r as its axis.

<u>Theorem 4:</u> [M27] Let $P \neq \phi$, dim $P < n$, and let x^k be a sequence generated by a relaxation algorithm.

(i) If $\lambda\epsilon(0,1)$, then x^k either terminates or else converges to a point of P.

(ii) If $\lambda = 2$, then x^k either terminates or else $\exists k_o$ such that, for $k \geq k_o$, x^k are on a spherical surface S_{n-r-1} having L_r as its center, where L_r is the minimum dimensional flat containing P. □

They also gave the first example of a nonterminating sequence. Consider Figure 2.1;the feasible region is defined by lines qq' and rr'.

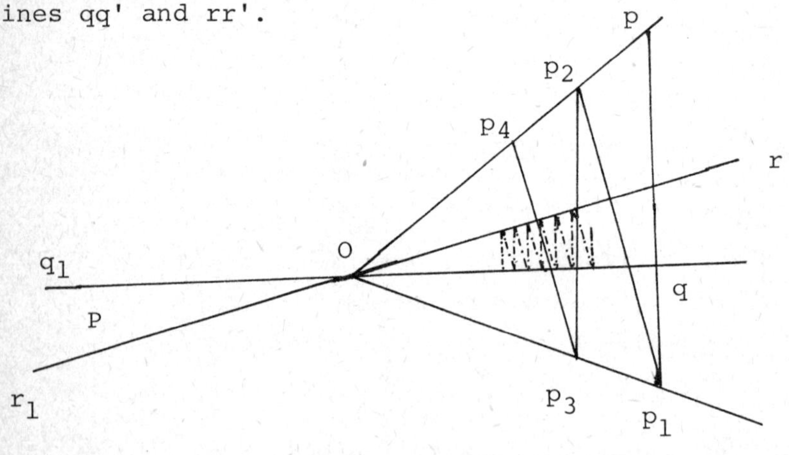

Fig. 2.1

Case for $\lambda = 1$ is shown in broken line. For $\lambda\in(0,1)$ it follows similarly. For $\lambda\in(1,2)$, it has the effect of changing $\vec{or}$ to $\vec{op}$, and similarly $\vec{oq}$ to $\vec{op_1}$ such that $pp_1/pq = \lambda$. Then applying for the new region $\lambda = 1$, we have the desired result.

We should point out that the example shows that given $\lambda\in(0,2)$, there exists a linear inequality system P with the required properties. It is quite possible that for $\lambda' > \lambda$, the relaxation algorithm might have finite convergence for λ' applied to problem P_λ. This point is made explicit by Goffin in [G29] .

Now let us consider a convex compact set P in R^n. At every point of its boundary we have a supporting hyperplane. Through these hyperplanes we can consider P as the solution set of an infinite family of linear inequalities. Then we have the following:

<u>Theorem 5:</u> [M27] Let P be as above and of dimension r, and let $P \subseteq L_r$, an r-flat. Suppose $x^o \notin P$, and x^k generated with $\lambda = 2$. Then:

(i) If $r = n$, then the process terminates.

(ii) For $r < n$, if $x^o \in L_r$, the process terminates. If $x^o \notin L_r$, then the infinite sequence eventually oscillates between two points which are symmetric with respect to L_r. □

Goffin studied the condition for which the relaxation method terminates in a finite number of iterations [G29]. A set C is called a <u>convex cone</u> if $x,y \in C$, and $\alpha \ge 0$, $\beta \ge 0$ implies that $\alpha x + \beta y \in C$, for all x,y,α,β. Given a finite index set I, and (unit) vectors $a_i \in R^n$, $i \in I$, we call $C = \{x \in R^n: \langle a_i, x \rangle \ge 0, \forall i \in I\}$ a <u>polyhedral cone</u>. For a cone C, let $C^* = \{u \in R^n: \langle u,x \rangle \le 0, \forall x \in C\}$ be its <u>dual cone</u>. Clearly when C is polyhedral, we have:

$$C^* = \{u \in R^n: u = - \sum_{i\in I} \alpha_i a_i,\ \alpha_i \ge 0,\ \forall i \in I\} .$$

For a unit vector e in R^n, and $\alpha\in[0,\pi]$, the <u>spherical cone</u> of axis e with half-aperture angle α is defined as:

$$C_\alpha(e) = \{x \in R^n: \langle x,e\rangle \geq ||x||\cos\alpha\}.$$

A cone C is <u>obtuse</u> if $-C^* \subset C$.

<u>Lemma 6:</u> [G29] A polyhedral cone C is obtuse iff any of the following holds:

(i) $a_i \in C,\ \forall i \in I$

(ii) The Gram matrix associated with a_i, $i \in I$, is non-negative, i.e. $[\langle a_i,a_j\rangle] \geq 0$

(iii) $\langle u,v\rangle \geq 0,\ \forall u \in C^*,\ v \in C^*$. □

For a polyhedral cone C, defined by a_i, $i \in I$, we define the condition number $\nu(C)$, as:

$$\nu(C) = \sup\{\sin\alpha: C_\alpha(e) \subseteq C,\ ||e|| = 1\ .$$

<u>Lemma 7:</u> $\nu = \max_{||u||\leq 1} \min_{i\in I} \langle a_i,u\rangle$ □

Given $P = \{x \in R^n: a_i x \leq b_i,\ i \in I\}$, with $||a_i|| = 1$, let us define $I(x) = \{i \in I: \ell_i(x) = f(x)\}$. Then the <u>tangent</u> or <u>supporting cone</u> to P at $y \in P$ is

$$C_P(y) = \{z \in R^n: \langle a_i,z\rangle \leq 0,\ i \in I(y)\},$$

and the <u>normal cone</u> is

$$N_P(y) = \{z: z = \sum_{i\in I} \alpha_i a_i,\ \alpha_i \geq 0\}.$$

A polyhedron P is obtuse at the boundary point y if

$$-N_P(y) \subset C_P(y).$$

A convex polyhedron is obtuse if it is obtuse at every boundary point, or equivalently at minimal faces; that is at vertices if it has vertices.

We define the condition number $\nu(P)$ of a polyhedron as

$$\nu(P) = \inf_{x\in\partial P} \nu(C_P(x)).$$

Lemma 8: [G29] A polyhedron P is full-dimensional if and only if $\nu(P) \geq 0$. P is obtuse if $\nu(P) \geq \sqrt{2}/2$. □

The following theorem generalizes Theorem 2.

Theorem 9: [G29] Suppose $P \neq \phi$ and $\lambda \in (0,2)$. Then the maximum distance relaxation algorithm either terminates or converges to a point in P. If it converges to x^*, then x^* is a boundary point. Moreover we have:

$$||x^k - x^*|| \leq 2d(x^o, P)\theta^k ,$$

and

$$d(x^k, P) \leq \theta^k \, d(x^o, P) ,$$

for some $\theta \in [0,1)$. □

The conditions for finite convergence are extended in the following theorem.

Theorem 10: [G29] Let P be full-dimensional. Then the maximum distance relaxation algorithm terminates finitely if

(i) P is obtuse and $\lambda \in [1,2]$, or

(ii) $\lambda \in [1,2]$, if $\nu(P) \geq \sqrt{2}/2$
or $\lambda \in (\lambda^*, 2]$, where

$$\lambda^* = 2/(1+2\nu(P)\sqrt{1-\;^2(P)}), \text{ if } \nu(P) < \frac{\sqrt{2}}{2} .$$

Furthermore if $\lambda > 2$, then the sequence either terminates finitely to a point of P or it does not converge. □

For recent studies on condition number of linear inequality system and its implications in relaxation methods see Mandel [M 5], Mangasarian [M10].

It is well-known that a linear program can be written as a system of linear inequalities: primal feasibility, dual feasibility, and reverse weak duality inequalities. A solution of the latter provides a pair of solutions for the primal and dual linear programs. One can apply a relaxation algorithm to solve the new system. This is observed by Oettli in [O 1]. Using a slightly more general formulation, he proved the geometric convergence of the algorithm. The version he considered is the classical projection method, i.e. $\lambda = 1$, and maximum distance. He also observed that the algorithm can be used to solve a pair of dual convex quadratic programs.

Hildreth [H] considered least square solution to a linear inequality system. In other words he wanted to solve the problem:

$$\min \ \frac{1}{2}||x||^2$$

$$\text{s.t.} \quad Ax \leq b.$$

Let $z^o \in R^m_+$ be any given point, $x^o = -A^T z^o$. Then a typical step:

$$x^{k+1} = x^k + c^k a_{ik},$$

$$z^{k+1} = z^k - c^k e_{ik},$$

$$c_k = \min \{z^k_{ik}, \lambda_k \frac{b_{ik} - a_{ik} x^k}{||a_{ik}||^2}\}$$

where i_k is determined cyclicly, $\lambda_k \in (\varepsilon_1 \ \varepsilon_2)$, $\varepsilon_1, \varepsilon_2 > 0$, $c^k \in R$.

The method is essentially a primal-dual method; z^k corresponds to dual variables. Geometrically, it is very similar to Agmon-Motzkin-Schoenberg method. Method is studied and applied by D'esopo [D19], Gorenflo and Kuvetz [G46], Wendler [W4], Oettli [O2], Herman and Lent [H14].

Bregman in 1965 [B28], gave a relaxation algorithm for finding a common point of a family of convex sets in a Hilbert space. Let A_i, $i \in I$, (not necessarily finite), be a closed convex set in the real Hilbert space H. Suppose $A = \bigcap_{i \in I} A_i$ is nonempty. Let $x^o \in H$, and let $i(x)$ be an index such that $x \notin A_i(x)$. Then:

$$x^{k+1} = P_{A_i}(x^k),$$

where $P_S(x)$ denotes projection of x onto convex set S, i.e. the nearest point of S to x.

We say x^k converges to x weakly, if $\forall u \in H$, $< u,x^k> \to < u,x >$. We will assume $x^* \in A$.

<u>Theorem 11:</u> [B 28]

(i) If I is finite then $x^k \to x^*$ weakly.

(ii) For any I, suppose $\forall x \in H$, $\max_{i \in I} d(x,A_i)$ exists, then $x^k \to x^*$ weakly.

(iii) When each A_i is a halfspace, at each point the maximum residue exists, and if $A_i = \{x: a_i x \le b_i\}$, $||a_i|| \le M$ for some M, then

$$x^n \to x^* \text{ weakly.}$$

Furthermore, when I is finite, the cyclic implementation also has the weak convergence property. □

Bergman also suggested a solution of the linear programming problem with the relaxation algorithm after transforming it to an inequality system by duality theory. A slightly different, but, in broader context, relaxation algorithm is earlier suggested by Bulavskii [B33] in 1962. Bergman also suggested a solution of convex quadratic programming with this algorithm. Furthermore he showed that one can use this algorithm to solve continuous linear programs (in function spaces).

He did not give any estimation of convergence rate. Interestingly he did not refer to any of the basic sources on relaxation methods. (See also [B34])

Censor and Lent [C 7 , C 8] extended Bregman method to a larger class of problems. They defined a class of functions, Bregman functions, which include norm function, and entropy function (i.e. $f(x) = 1/2||x||^2$, and $f(x)= \Sigma x_i \log x_i$, for $x > 0$). For a Bregman function f, $f \in C'$, they defined a function $D: \bar{S} \times S \to R$ as

$$D(x,y) = f(x)-f(y) - <\nabla f(y), x-y> \quad ,$$

where S is the domain of f, $\bar{S}$ is the closure of S. Let $Q \in R^n$, a point $\bar{x} \in Q \cap \bar{S}$ for which

$$\min_{z \in Q \cap \bar{S}} D(z,y) = D(\bar{x},y)$$

is called D-projection of y onto Q, and denoted by $P_Q(y)$. Then this projection is replaced by the usual orthogonal projection and selection of row i is either determined by largest $D(x,x^k)$ among the sets Q or cyclicly.

Merzlyakov [M17], in 1962 extended the basic Agmon-Motzkin-Schoenberg algorithm in a way allowing one to combine several violated constraints by taking a positive linear combination of the defining normals. He also proved the geometric convergence and gave an explicit formula for the convergence rate.

Now let us consider the problem of finding a feasible point to the following system of convex inequalities:

$$f_i(x) \leq 0, \quad i = 1,\ldots,m ,$$

where each $f_i: R^n \to R$ is a convex function.

Let $M_i = \{x: f_i(x) \leq 0\}$ and $M = \bigcap_{i=1}^{m} M_i$. Suppose that $M \neq \phi$.

A basic fact used inthe relaxation algorithm for a linear inequality system is that each hyperplane is a seperating hyperplane for the solution set. This property can easily be captured by approximating the convex function by a linear (affine) function at a point using the subgradient inequality. Even though we con no longer guarantee the next projected point with $\lambda = 1$ to be feasible with respect to M_i , we still have Fejer-monotonicity with respect to solution set M or M_i. Thus with this interpretation one can apply the standard relaxation algorithm to the above problem. Eremin, in a series of papers [E19,18,19], did this and more, and proved the convergence for $\lambda \in (0,2)$. He also applied this approach to general convex programming via a penalty function technique [E16].

Eremin in [E15]calculated the rate of convergence of the relaxation method for finding a solution to a system of convex inequalities.

Consider the system of inequalities:

$$f_i(x) \le 0, \quad i = 1,\dots,m.$$

He assumes a regularity condition, namely the Slater constraint condition, that is, $\exists\ \bar{x}$ s.t. $f_i(\bar{x}) < 0, \ \forall i$.

One can replace the above system with a single function in many ways. Eremin considers:

(i) $s_1(x) = \max_i f_i(x)$

(ii) $s_2(x) = \sum_i K_i f_i^+(x)$,

where $f_i^+(x) = \max\{f_i(x), 0\}$, and $K_i \ge 0$.

He proves that the method converges with geometric rate $\theta \epsilon (0,1)$, where

$$\theta^2 = 1-\lambda(2-\lambda)\ \frac{1}{C^2K^2},$$

where K is an upper bound on the norms of subgradients, and C measures some condition number of the subgradients of the system.

Eremin and Mazuzov [E20], consider the following programming problem:

$$\min \ < c,x >$$
$$\text{s.t.} \quad f_i(x) \le 0, \quad i = 1,\dots, .$$

Let us define $M = \{x: f_i(x) \le 0\}$. Let us assume existence of functions $d(x) \in R$, and $e(x) \in R^n$, such that:

(i) $M = \{x: d(x) \le 0\}$,

(ii) For $y \notin M$, $< e(y),\ x-y > + d(y) \le 0, \ \forall x \in M$, and $||e(y)|| = 1, \ \forall y \in M$.

Several examples of d(x) are well-known; e(y) represents a (normalized) seperating hyperplane of M at y. Then:

$$\Phi(x) = x - \lambda d(x)e(x)$$

is a typical relaxation type iteration for $\lambda\in(0,2)$ and $x \notin M$. It is well-known that Φ is Fejer-monotone with respect to M.

They suggested the following method:

$$x^{k+1} = \Phi^{n_k}(x^k) - \lambda_k c,$$

where n_k is a natural number satisfying

$$d[\Phi^{n_k}(x^k)] \le \lambda_k,$$

for some $\lambda_k \in (0,\lambda_o]$, $\lambda_o > 0$.

Let $S(\varepsilon) = \{x| < c,x > - < c,x^* > \le \varepsilon, d(x) \le \varepsilon|$, $\varepsilon > 0$, $\delta(\varepsilon) = \sup_{x\in S(\varepsilon)} || x-M^*||$, $D(\varepsilon) = \delta(\varepsilon) + 2\varepsilon$. Note $x^* \in M^*$ represents an optimal solution. They proved the following result:

<u>Theorem 12:</u> [E 20] Suppose $M^* \ne \Phi$ and is bounded. For the above algorithm we have:

(i) if $\Sigma\lambda_k = \infty$ then $\exists N_o$ s.t. $k \ge N_o$, $||x^k - M|| \le D(\varepsilon)$, and $|< c,x^k > - < c,x^* >| \le D(\varepsilon)$;

(ii) if $\lambda_k \to 0$, $\Sigma\lambda_k = +\infty$, then $d(x^k,M^*) \to 0$, and consequently $< c,x > \to < c,x^* >$. □

In relaxation methods we treated the feasibility problem and always assumed that the problem is feasible. It is well-known that any optimization problem can be reduced to, at least a sequence of feasibility problems. But then we can no longer assume feasibility. We have to change the algorithm: the result is a subgradient algorithm.

Subgradient algorithms, as we know today, started with Shor's work on a flow problem in 1962 [S 9]. Shor gave a full account in his dissertation in 1964 [S10].

At the same time Eremin, in his work on Chebysev solution of an incompatible linear inequality system, suggested a subgradient algorithm in 1962 [E 13].

Let $f: R^n \to R$ be a convex function, Let $\partial f(x)$ be the set of all subgradients of f at x, i.e.

$$\partial f(x) = \{u \in R^n : f(y)-f(x) \geq \langle u, y-x \rangle, \quad \forall y \in R^n\}.$$

Then a typical iteration of the subgradient algorithm is

$$x^+ = x - \alpha \frac{u}{||u||}, \quad u \in \partial f(x), \quad \alpha > 0.$$

Convergence of the algorithm depends on the step size. Step size is determined, without any line search, with apriori information, sometimes depending on $f(x^*)$, the optimal value.

Defining $f_1(x) = \max_i \{a_i x - b_i\}$ and $f(x) = [f_1(x)]^+ = \max\{f_1(x), 0\}$, we can realize a solution of the linear inequality system as a convex minimization problem. Furthermore, since by assumption $P = \{x: Ax \leq b\} \neq \phi$, we have $f(x^*) = 0$. The fact that $u \in \partial f(x)$ identifies the method as the maximum residue version of the relaxation algorithm with λ, where $\lambda f(x) = \alpha$. In general, if we know $f(x^*)$, but do not require $f(x^*) = 0$, then λ must satisfy the condition $\alpha(f(x)-f(x^*)) = \lambda$. Thus only a version of Agmon-Motzkin-Schoenberg method realizes as a subgradient method.

On the other hand, since any convex function can be written as a (possibly infinite) linear inequality system through the subgradient inequalities written at each point of the domain, the subgradient algorithm can be realized as a maximum distance relaxation algorithm.

One basic feature of subgradient optimization is the following: even though the function values do not decrease at every iteration, the distance from the optimum set decreases if proper step size is chosen. However any subgradient is a descent direction for the distance function for this set (for a lucid proof of this, see Wolfe [W 6]).

Eremin [E13], Poljak [P12,P13], and Ermelov [E24] showed that when the step size λ_j, at iteration j satisfies the following conditions:

$$\lambda_i \geq 0, \quad \lim \lambda_i \to 0, \quad \Sigma \lambda_i = \infty,$$

the method converges.

When estimate $\hat{f}$ of $f(x^*)$ is available the iteration takes the form

$$x^+ = x - \alpha \frac{f(x)-\hat{f}}{||u||^2} u, \quad u \in \partial f(x).$$

This form is used by several researchers [P13,O1 ,H10].

Held and Karp [H9], in their pionering work on traveling salesman, unaware of Soviet literature, rediscovered the subgradient algorithm via solution of a large-scale, combinatorially defined linear program. They suggested the following iteration

$$x^+ = x - \lambda u, \quad u \in \partial f(x).$$

Shor [S], suggested various forms:

(i) $x^+ = x - \alpha \frac{u}{|u|}$, $u \in \partial f(x)$,

and λ is constant each iteration,

(ii) $\lambda_k = K \rho^k$,

where $\rho < 1$ depends on a certain condition number of the convex function (we will study this version in Chapter 3). He showed the geometric convergence of the sequence. The above step size rule is analyzed by Goffin as a function of ρ in relation to the "condition number of the convex function and K" [G26].

Camerini et al [C 4], suggested to use a modified direction which is obtained by adding a multiple of the previous modified direction to a subgradient at the current point, provided that their inner product is nonpositive.

In this regard, the relationship between maximum distance relaxation and subgradient optimization is studied by Goffin in [G27]. He showed that, under certain conditions, the sustainable rate of convergence for subgradient optimization is exactly equal to the rate of relaxation with $\lambda = 1$.

Poljak [P 13], extensively studied convergence of relaxation-subgradient algorithms for a variety of convex problems in finite and infinite dimension. The convergence rate almost always is geometric. For example, when f is strongly convex i.e.

$$f(x)-f(x^*) \geq m||x-x^*||^2 ,$$

and $f'(x)$ is Lipschitz with M, then for $\lambda\varepsilon(\varepsilon_1, 2-\varepsilon_2)$, one has a geometric convergence rate:

$$q = \sqrt{1-\varepsilon_1\varepsilon_2 \frac{m^2}{M^2}} .$$

M can be replaced with an upper bound on the norm of (sub) gradients on a compact set.

There has been some effort to test the subgradient-relaxation algorithms emprically. The most notable are Held-Wolfe-Crowder [H 11], Held-Karp-Wolfe [H 10], Crowder [C 35], Goffin [G 28, 30, 31, 34], Camerini et al [C 4]. Results are favourable. One negative result came from an experiment in solving certain linear complementarity problems via linear programming [C 33]. Relaxation-subgradient methods are widely used in large-scale linear programming and in combinatorial programming via Lagrangian duality. Some references are Shapiro [Z14, Z15], Dantzig et al [D 11], Fisher [F 6], Burkard [B 35], and Sandi [S 8].

For very recent Soviet research on relaxation-subgradient methods: see Ivanov [I 4], Kreinin [K23], Poljak [P17], Bulavskii [B34], Eremin [E21-23], Demjanov [D18].

It is known that for solving shortest path problems, the dual simplex method is equivalent to a relaxation algorithm (see Lawler [L 2 ,p.79]). In this sense, the relaxation methods are often called dual methods.

It should be noted that there are some other relaxation methods for solving, in particular, quadratic programming problems [C 36,S 4]. They differ from the ones we studied, in the sense that they use second order information; hence they need second order differentiability. However they enjoy the same geometric convergence.

For a survey of the Soviet research on rather general relaxation processes in a more abstract setting see Lyubich and Maistrovskii [L 19].

2.2 TOTALLY UNIMODULAR MATRICES

A matrix $A \in R^{m \times n}$ is said to be totally unimodular (t.u.) if for every square submatrix B of A we have det $B \in \{0,1,-1\}$. The following properties are easy consequences of the definition, and are well-known.

Lemma 1:

(i) A is t.u. <=> A^t is t.u.

(ii) A is t.u. <=> every square submatrix of A is t.u.

(iii) $A \in R^{m \times n}$ is t.u. <=> $[A, I_m]$ is t.u.

(iv) Let B be obtained from A by multiplying a row of A by -1. Then A is t.u. <=> B is t.u.

(v) Let $[B, I_n]$ be obtained from A by a sequence of elementary row operations. If A is t.u. then $[B, I_n]$ is t.u.

(vi) For a nonsingular t.u. A we have

$$A^{-1}_{ij} \in \{0, \pm 1\} . \qquad \square$$

The importance of totally unimodular matrices lies in the fact that the vertices of the following polyhedron:

$$\underline{b} \le Ax \le \bar{b},$$

$$\underline{d} \le x \le \bar{d} ,$$

are integer valued, whenever $\underline{b}$, $\bar{b}$, $\underline{d}$, $\bar{d}$ are integer valued ($\pm \infty$ are allowed). This is the celebrated result of Hoffman and Kruskal [H19] (see also Veinott and Dantzig [V 2]).

Theorem 2: [H19,V2] If A is an integral matrix, the following are equivalent.

(i) A is totally unimodular

(ii) The extreme points of $\{x: Ax \le b,\ x \ge 0\}$ are integral for all integral b

(iii) Every nonsingular submatrix of A has an integral inverse. $\square$

Thus for totally unimodular A and integral b, c, the following LP and associated dual LP have integral solutions:

$$\begin{aligned} \max:\ & cx \\ \text{s.t.}\ & Ax \le b, \\ & x \ge 0. \end{aligned}$$

Necessary and sufficient conditions for total unimodularity have been studied for a long time by various researchers. See for example Hoffman [H 20], Padberg [P 1], Truemper [T 8-9].

Recently P.D. Seymour [S 6] obtained a decomposition theory for totally unimodular matrices. It is done in the spirit of matroid theory. It says that any t.u. matrix can be constructed by piecing together network matrices and copies of a certain ten element matroid.

2.3 THE ALGORITHM: VERTEX

Dantzig [D 6 , p.113, problems 22,23] suggested that given a non-basic feasible solution to a linear program in standard equality form, a simplex method can be used by adjusting the values of non-basic variables to find a basic feasible solution (vertex) with better objective function.

Independently and at the same time Charnes et al [C 10,C 11] suggested a similar algorithm for the same purpose.

Maurras [M 12,M 13], first suggested to use this algorithm to find an optimal (basic) solution from an approximate solution.

In order to be complete, we give a description of the algorithm, and hence a proof of its polynomiality. Furthermore we use the algorithm in Chapter 4 in solving some hard linear programming problems.

Given a feasible $\bar{x}$ of following system:

$$\begin{aligned} & A\bar{x} = b \\ & 0 \le \bar{x} \le d\ , \end{aligned}$$

the algorithm VERTEX finds a vertex $\tilde{x}$ such that $c\tilde{x} \ge c\bar{x}$. Let us define a maximal independent set J as basis of A. W.l.o.g. we may assume our system has the form:

$$[A_1 | I]\ \bar{x} = b, \quad c = (c_{\bar{J}},\ c_J),$$

where $|J| = \text{rank}(I)$, $c_J = 0$, $J \cup \bar{J} = \{1,2,\ldots,n\}$.

Most of the time we have a system $Ax \leq b$. Hence the above form is readily available. Otherwise one can obtain (A_1,I) in n pivots. Let

$$\tilde{J}(x) = \{j \notin J(x) : 0 < x_j < d_j\},$$

where $J(x) = \{j_1,j_2,\ldots,j_m\}$ denotes the set of indices of the basis. If $\tilde{J}(x) = \phi$, then x is a basic feasible solution. Otherwise we will pivot and thus decrease the cardinality of $\bar{J}$. Let us specify the pivoting: for $\ell \in \tilde{J}(x)$, let

$$(*) \quad \begin{cases} x_{j_i}(t) = x_{j_i} - \sigma_\ell \cdot t a_{i\ell} & i = 1,\ldots,m\ , \\ x_\ell(t) = x_\ell + \sigma_\ell \cdot t\ , & \\ x_k(t) \equiv x_k\ , & k \in \tilde{J}(x) \setminus \ell \end{cases}$$

$$\text{where } \sigma_\ell = \text{sing}(c_\ell) = \begin{cases} -1 & \text{if} \quad c_\ell \leq 0 \\ +1 & \text{if} \quad c_\ell > 0 \end{cases} .$$

We will represent (*) as

$$x(t) = x + t(\sigma_\ell\ \tilde{a}_\ell),$$

$$\text{where } (\tilde{a}_\ell)_k = \begin{cases} a_{i\ell} & \text{for} \quad k = j_i \\ 1 & \qquad k = \ell \\ 0 & \qquad \text{otherwise.} \end{cases}$$

Now t will be determined by (*) such that one component of x among $\{j \in J \cup \ell\}$ becomes zero or reaches its upper bound. If ℓ reaches one of its bounds, then $\tilde{J} = \tilde{J} - \{\ell\}$. If some $j \in J$ reaches one of its bounds then $J = J \cup \{\ell\} - \{j\}$, $\tilde{J} = \tilde{J} - \{\ell\}$. If more than one $j \in J$ reaches its bounds, delete them from J and extend J to a basis using columns in $\tilde{J}$.

Thus at each iteration $|\tilde{J}(x)|$ decreases by (at least) one. So in at most n iterations $\tilde{J}(x) = \phi$, where x is a vertex.

For any selection rule for $\ell \in \tilde{J}(x)$ one reaches a vertex with the required properties. We implemented the following rule in our computations in Chapter 4. Let $\ell \in \tilde{J}(x)$ such that

$$\max_{i \in \tilde{J}(x)} |c_i| = |c_\ell| .$$

In the absence of an objective function the selection rule simplifies to: for $j \in \tilde{J}(x)$, let $\sigma_j = -1$.

A different approach to the algorithm is by means of the projection method. If one applies the projection method starting at any feasible point, in at most n iterations one reaches a vertex.

2.4 BASIC ALGORITHMS

Agmon [A 3] conjectured that when $x^+ = x' + \gamma a_i$, for small positive γ, i.e. overrelaxation, convergence might be faster.

R.G. Jeroslow [J 1], and independently J.F. Maurras [M13], suggested the following modification of the Agmon-Moltzkin-Schoenberg algorithm. (Notation: from now on, in this chapter, x^k denotes a vector, and x_i denotes i'th component of x.)

(1) Pick $x^o \in R^n$, set $k = 0$

(2) If $\ell_i(x) \leq 0$, $\forall i$, stop.

(3) Otherwise choose i such that $\ell_i(x) > 0$, and set

$$x^{k+1} = x^k - \frac{a_i}{||a_i||} \left(\frac{\ell_i(x)}{||a_i||} + \varepsilon\right)$$

(4) $k = k + 1$, go to 2.

(1)

Let $P_\varepsilon = \{x \in R^n\colon a_i x \leq b_i - \varepsilon, \forall i\}$, $M \geq ||x_o - y||$ for some $y \in P_\varepsilon$, and $P = \{x \in R^n\colon a_i x \leq b_i, \forall i\}$ $= \{x\colon \ell_i(x) \leq 0, \forall i\}$.

Theorem 1: The above algorithm finds a solution (in P) in at most M^2/ε^2 iterations, or proves that $P_\varepsilon = \phi$. □

We need a basic lemma.

Lemma 2: Let $x^i \notin P$, $y \in P_\varepsilon$, $||x^i - \tilde{x}^i|| = d$. Then

$$||x^i - y||^2 - ||x^{i+1} - y||^2 \geq (d + \varepsilon)^2. \qquad (2)$$

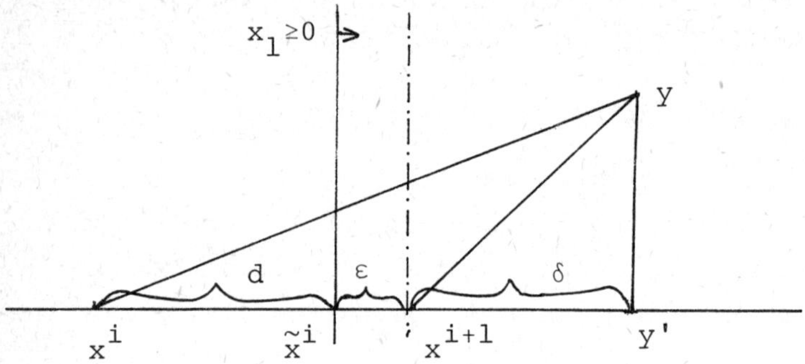

Fig. 2.2

Proof: Without loss of generality we can assume that the hyperplane $\ell_i(x) \geq 0$ is $e_1 x \geq 0$, if necessary by rotating and translating the space, or by coordinate changes by orthogonal transformation. We have

$$||x^i - y||^2 = ||-y + y'||^2 + ||x^i - y'||^2 ,$$

where y' is the projection of y on $\{te_1 : t \in R\}$, e_1 is the usual unit vector. Similarly

$$||x^{i+1} - y||^2 = ||x^{i+1} - y'||^2 + ||y - y'||^2 .$$

Thus

$$||x^i - y||^2 - ||x^{i+1} - y||^2 = ||x^i - y'||^2 - ||x^{i+1} - y'||^2 .$$

Since $y \in P_\varepsilon$, $\ell_i(y') \geq \varepsilon \Rightarrow \delta = y_1' - x^{i+1} \geq 0$.
Clearly

$$||x^i - y||^2 - ||x^{i+1} - y||^2 = (\delta + d + \varepsilon)^2 - \delta^2 \geq (d + \varepsilon)^2 . \qquad \square$$

Proof of Theorem: Suppose $P_\varepsilon \neq \phi$, and $y \in P_\varepsilon$, and $M \geq ||x^o - y||$. By Lemma 2 $g_i = ||x^i - y||^2$ gets smaller by at least ε^2 (in fact $(d + \varepsilon)^2$) at each iteration. Hence $K = M^2/\varepsilon^2$ iterations later we have $||x^K - y||^2 < 0$, a contradiction. $\square$

Remark: Let $\alpha = ||x^i - y||$, $\alpha - \beta = ||x^{i+1} - y||$.
Then by Lemma 2, we have

$$\alpha^2 - (\alpha - \beta)^2 \geq (d + \varepsilon)^2 \geq \varepsilon^2 .$$

Since

$$2\alpha\beta \geq 2\alpha\beta - \beta^2 \geq \varepsilon^2 ,$$

we have

$$\beta \geq \frac{\varepsilon^2}{2\alpha} \geq \frac{\varepsilon^2}{2M} .$$

Then $K = 2M^2/\varepsilon^2$ is an upper bound for the above algorithm, as given by Maurras [M13].

Because of its importance we will give a second proof of the above lemma.

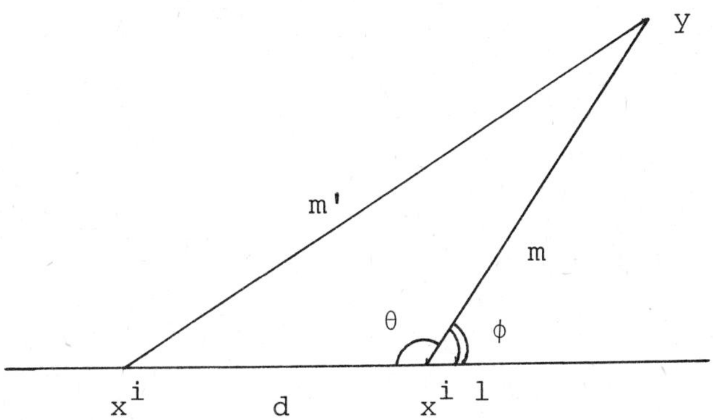

Fig. 2.3

Let $||x^i - y|| = m'$, $||x^{i+1} - y|| = m$, $||x^i - x^{i+1}|| = d$.

Let us note that we are no longer assuming that $x^{i+1} - x^i = \alpha a_{i'}$ for some α and i'. The only thing we need is that we get closer to the polyhedron, or equivalently $m' > m$, $d > 0$, $(d \geq \varepsilon)$. Since the points x^i, x^{i+1}, y form a triangle in R^n, we can talk about angles θ , ϕ (see Fig. 2.3).

Using the law of cosines we can write

$$m^2 + d^2 - 2dm \cos \theta = m'^2 .$$

Solving the above quadratic equation for m, we have

$$m = d \cos \theta \pm \sqrt{m'^2 - d^2(1-\cos^2 \;)}$$

$$= d \cos \theta \pm \sqrt{m'^2 - d^2 \sin^2 \theta} .$$

Assuming in general $m' >> d$, we have

$$m = d \cos \theta + \sqrt{m'^2 - d^2 \sin^2 \theta} \leq m' + d \cos \theta - \frac{d^2 \sin^2 \theta}{2 m'}$$

(since, by Taylor's Theorem,

$$\sqrt{m^2 - x^2} \leq m - \frac{x^2}{2m}) \quad ,$$

which implies

$$m' - m \geq - d \cos \theta + \frac{d^2 \sin^2 \theta}{2m'} .$$

Note that $\theta + \phi = \pi$, hence we have $- \cos \theta = \cos \phi$ and $\sin \theta = \sin \phi$. Therefore, we have

$$m' - m \geq d \cos \phi + \frac{d^2 \sin^2 \phi}{2 m'} .$$

For $\phi \in [0, \pi/2]$:

$$d \cos \phi + \frac{d^2 \sin^2 \phi}{2\, m'} \geq \frac{d^2}{2\, m'} \geq \frac{\varepsilon^2}{2\, m'} \geq \frac{\varepsilon^2}{2M}$$

which also gives the bound $K = 2M^2/\varepsilon^2$.

The importance of the above result lies in the observation that if one could bound ϕ away from $\pi/2$, say $\phi \leq \phi_o < \pi/2$, then the bound in the theorem will reduce to $1/\cos \phi_o \cdot M/\varepsilon$, a linear function instead of quadratic in (M/ε).

Suppose we also have some simple constraints in the form

$$0 \leq x \leq d .$$

Let $T = \{x: 0 \leq x \leq d\}$. Then for $x \notin T$, we define

$$\pi(x) = \pi_T(x) = \begin{cases} 0 & x_j < 0 \\ d_j & x_j > d_j \\ x_j & \text{otherwise} \end{cases} \tag{3}$$

$\pi(x)$ is the projection of x onto T. Clearly, for $y \in T$ we have

$$||\pi(x) - y|| \leq ||x - y|| , \tag{4}$$

the validity of which can be veried by the successive application of Lemma 2. If some of the restrictions in the definition of T are removed, a similar $\pi(x)$ can be defined, with obvious properties naturally.

One may wonder whether a more careful analysis could improve the factor M^2/ε^2 in A0. The following example shows that this cannot be done without a change of A0. Suppose we have just two constraints in the plane as indicated in Fig. 3. Iterations $k = 1,2,3,\ldots$ of Algorithm A0 result in indicated points x^1, x^2, x^3, when $x^o = (\varepsilon/\sin\theta, 0)$. Assume we stop the algorithm when x_1-component of the current solution becomes nonpositive for the first time. The number of elapsed iterations, N, is certainly a lower bound on the performance of the algorithm. Since $\eta_o = 4\varepsilon \sin\theta$, and $\eta_o \geq \eta_k$, $\forall k \leq N$, we have

$$N \geq \frac{x_1}{4\varepsilon \sin\theta} = \frac{x_1^o}{4\varepsilon^2} = \frac{(x_1^o)^2}{4\varepsilon^2} = \frac{M^2}{16\varepsilon^2}$$

if we choose $M = 2x_1^o$, the minimal distance between x^o and Q_e

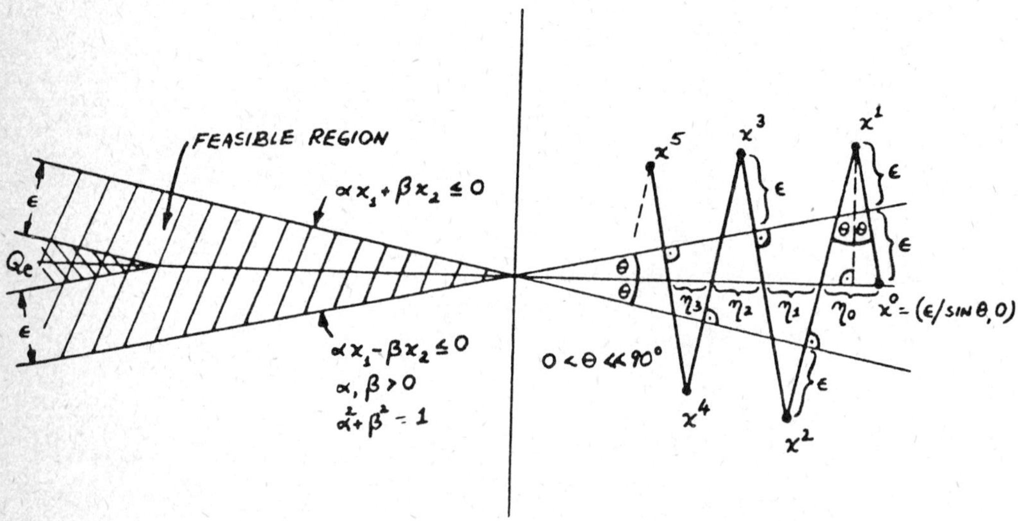

Fig. 2.4

To be complete, we will present the above algorithm as A0 in the sprit of upcoming algorithms.

Algorithm A0

Input:

1) System

$$Ax \le b,$$
$$0 \le x \le d, \tag{5}$$

where for each row a_i of $m \times n$ matrix A, $||a_i|| = 1$,

2) $\varepsilon > 0$, let $e = (\varepsilon, \varepsilon, \ldots, \varepsilon)^t$

3) Positive M such that $M \ge ||y||$ for some

$$y \in Q_e = \{x \; Ax \le b - e, \quad 0 \le x \le d\}, \text{ if } Q_e \ne \phi$$

Output. "$Q_e = \phi$" or x^* satisfying (5).

Complexity. $O(mn\, M^2/\varepsilon^2)$

Procedure.

0. Set $g_o = M^2$, $r = 0$, $x^o = 0$.
1. If $a_i x^r \le b_i$, $\forall i$, let $x^* = x^r$, and stop.
2. If $g_r \le 0$, write "$Q_e = \phi$" and stop.
3. Select any i such that $h_r = a_i x^r - b_i > 0$, and let

$$\bar{x} = x^r - (\varepsilon + h_r) a_i ,$$
$$\bar{\bar{x}} = \pi(\bar{x}),$$
$$g_{r+1} = g_r - \{(\varepsilon + h_r)^2 + ||\bar{x} - \bar{\bar{x}}||^2 ;$$

go to 1. □

Validity of the algorithm is proved earlier. The algorithm iterates at most M^2/ε^2, and since the complexity of each iteration is $O(mn)$, we obtain the given complexity.

We will very often encounter the problem of finding a feasible solution of the system

$$\begin{aligned} cx &\geq v \\ Ax &\leq b \\ 0 \leq x &\leq d\ . \end{aligned} \tag{6}$$

We will give a modified version of A0, to handle the objective function constraint directly.

Algorithm A1.

Input:

1) System (6)

where for each row a_i of $m \times n$ A, $||a_i|| = 1$

2) $\varepsilon > 0$; let $e = (\varepsilon, \varepsilon, \ldots, \varepsilon)^t$.

3) Positive M such that $M \geq |y|$ for some $y \in Q_e = \{x | cx \geq v,$ $Ax \leq b - e,\ 0 \leq x \leq d\}$ if $Q_e \neq \phi$.

Output: "$Q_e = \phi$" or x^* satisfying (6).

Complexity: $O(\bar{m} \cdot n \cdot M^2/\varepsilon^2)$, where $\bar{m} = \max\{m, \log n\}$.

Procedure:

0. Set $g_o = M^2$, $r = 0$, and $x^o = \bar{\bar{x}} = 0$. Go to 4.
1. If $a_i x^r \leq b_i$, $\forall i$, let $x^* = x^r$ and stop.
2. If $g_r \leq 0$, write "$Q_e = \phi$" and stop.
3. Select any i such that $h_r = a_i x^r - b_i > 0$, and compute $\bar{x} = x^r - (\varepsilon + h_r) a_i^t$.

   Then define $\bar{\bar{x}}$ by

$$\bar{\bar{x}} = \pi(\bar{x})\ .$$

Let $g_{r+1} = g_r - \{(\varepsilon + h_r)^2 + ||\bar{x} - x||^2\}$.

4. If $c\bar{\bar{x}} \geq v$, let $x^{r+1} = \bar{\bar{x}}$, replace r by r + 1, and go to 1.

5. Define $R = \{j | \bar{\bar{x}}_j = 0,\ c_j < 0\}$ and $S = \{j | \bar{\bar{x}}_j = d_j,\ c_j > 0\}$. If $c_j = 0$, $\forall j \notin R \cup S$, write "$Q_e = \phi$" and stop. Else find largest $p > 0$ (by ratio test) such that $\tilde{x}$, defined by

$$\tilde{x}_j = \begin{cases} \bar{\bar{x}}_j + pc_j & \text{if} \quad j \notin R \cup S \\ \bar{\bar{x}}_j & \text{otherwise,} \end{cases}$$

satisfies $0 \leq \tilde{x} \leq d$ and $c\tilde{x} \leq v$. Replace $\bar{\bar{x}}$ by $\tilde{x}$ and go to 4.

<u>Proof of validity:</u> If v is small enough so that the algorithm always branches to step 1 from step 4, then A1 is identical to A0. If it branches to step 5, and if for some $y \in Q_e$ such that $g_{r+1} \geq ||\bar{\bar{x}} - y||^2$, we have $||\tilde{x} - y|| < ||\bar{\bar{x}} - y||$, then $g_{r+1} \geq ||\tilde{x} - y||^2$, and we have the validity of algorithm.

So we now assume that $c\bar{\bar{x}} < v$ occurs in step 4, so step 5 is entered. If $c_j = 0$, $\forall j \notin R \cup S$, then $c\bar{\bar{x}} = \max\{cx | 0 \leq x \leq d\}$, and Q_e must be empty. Else take any $y \in Q_e$ such that $g_{r+1} \geq ||\bar{\bar{x}} - y||^2$. We will show that $||\tilde{x} - y|| < ||\bar{\bar{x}} - y||$, where $\tilde{x}$ is the vector specified at the end of step 5. Define $\hat{c}$ and $\hat{y}$ by

$$\hat{c}_j = \begin{cases} 0 & \text{if} \quad j \in R \cup S \\ c_j & \text{otherwise} \end{cases}$$

$$\hat{y}_j = \begin{cases} 0 & \text{if} \quad j \in R \\ d_j & \text{if} \quad j \in S \\ y_j & \text{otherwise} \end{cases} .$$

By the above assumptions and definitions $c\bar{\bar{x}} < v \le cy \le c\hat{y}$, hence $\hat{c}\bar{\bar{x}} < \hat{c}\hat{y}$. Vector $\tilde{x}$ satisfies $\tilde{x} = \bar{\bar{x}} + p(\hat{c})^t$, $0 \le \tilde{x} \le d$, $c\tilde{x} \le v$ for some $p > 0$. Since $c\tilde{x} > c\bar{\bar{x}}$ and $\tilde{x}_j = \bar{\bar{x}}_j$, $\forall j \in R \cup S$, we have $\hat{c}\bar{\bar{x}} < \hat{c}\tilde{x} \le \hat{c}\hat{y}$. But this is essentially an instance of the previously described process involving one inequality, in the subspace $X = \{x_j = 0,\ j \in R \cup S\}$. Since $\tilde{x} = \bar{\bar{x}} + pc$, we have $||\bar{\bar{x}} - \hat{y}||^2 - ||\tilde{x} - \hat{y}||^2 > 0$ by lemma 2. (It is obviously irrelevant that $||\hat{c}||$ may not be 1). But then $||\bar{\bar{x}} - y||^2 - ||\tilde{x} - \hat{y}||^2 = ||\bar{\bar{x}} - \hat{y}||^2 - ||\tilde{x} - \hat{y}||^2 > 0$, and g_{r+1} is an upper bound on $||\tilde{x} - y||^2$. In fact one can update g_{r+1}, as $g_{r+1} = g_{r+1} - ||\tilde{x} - \bar{\bar{x}}||^2 = g_{r+1} - p^2||\hat{c}||^2$. Finally we note that step 5 may be executed at most n times while index r remains unchanged, and by induction $g_{r+1} \ge ||x^{r+1} - y||^2$. Let us recall that sorting will take $O(n \log_2 n)$ time [A 4]. Proper implementation results in $O(n \log_2 n)$ complexity for these iterations with constant r, so the overall effort is $O(\bar{m} \cdot n \cdot M^2/\varepsilon^2)$, where $\bar{m} = \max\{m, \log_2 n\}$. □

Finally it should be noted that a trivial modification allows one to process systems where $\underline{b} \le Ax \le \bar{b}$ takes the place of $Ax \le b$, and where some x_j have no upper and/or lower bound. Set Q_e and bound M are adjusted in the obvious way, and the complexity formula remains valid.

We now construct algorithms that use A1 as subroutine.

<u>Algorithm A2</u>.

<u>Input</u>:

1. Linear program

$$\begin{aligned} \max:\ & cx \\ \text{s.t.}\ & Ax \le b \\ & 0 \le x \le d, \end{aligned} \tag{7}$$

where A is $m \times n$ and $||a_i|| = 1$, $\forall i$.

2. $\varepsilon > 0$, $\delta > 0$; let $e = (\varepsilon, \varepsilon, \dots, \varepsilon)^t$.

3. Positive $M \geq ||\bar{x}||$, $U \geq c\bar{x}$, $K \geq U - c\bar{\bar{x}}$, where $\bar{x}$ is optimal for

$$\begin{aligned} \max: \quad & cx \\ \text{s.t.} \quad & Ax \leq b - e \\ & 0 \leq x \leq d\,, \end{aligned} \tag{8}$$

and $\bar{\bar{x}}$ is a feasible solution of (8).

Assumption: $P_e = \{x | Ax \leq b - e,\ 0 \leq x \leq d\} \neq \phi$

Output: Feasible x^* of (7) satisfying $cx^* \geq c\bar{x} - \delta$.

Complexity: $O(\bar{m}\cdot n\cdot M^2\cdot N/\varepsilon^2)$ where $\bar{m} = \max\{m, \log_2 n\}$ and $N = \max\{\log_2(K/\delta), 1\}$.

Procedure: Do dichotomous search for $v^* \in [U-K, U]$ such that Algorithm A1

(i) terminates with a solution to $cx \geq v^*$, $Ax \leq b$, $0 \leq x \leq d$ (let x^* be this solution), or

(ii) proves that $cx \geq v^* + \delta$, $Ax \leq b - e$, $0 \leq x \leq d$ has no solution.

Parameters ε and M for A1 are those of the input.

Proof of validity: First note that assumption "$P_e \neq \phi$" implies that all parameters of the input are well defined. Suppose we apply A1 to (6) using the given ε and M. If $v \leq c\bar{x}$, then $\bar{x} \in Q_e$, and $M \geq ||\bar{x}||$ implies that A1 must terminate with a solution to (6). By dichotomous search we obtain the desired v^* and x^* with $O(\log_2(K/\delta))$ applications of A1. Hence total effort is of order $O(\bar{m}\cdot n\cdot M^2\cdot \log_2(K/\delta)/\varepsilon^2)$ provided $K \geq \delta$. But if $K < \delta$, A1 must be applied just once. The formula for N combines both cases. □

A minor modification of A2 is needed if some bounds of type $x_j \geq 0$ and/or $x_j \leq d_j$ are removed from (7). In that case one should add the assumption "(7) has an optimal solution".

Algorithm A3.

Input:

1. Linear program (7).
2. $\varepsilon > 0$.
3. Positive $M \geq \max\{||\bar{x}||, ||\hat{x}||\}$, $U \geq c\bar{x}$, $K \geq U - c\bar{\bar{x}}$, where $\hat{x}$ is optimal for (7), $\bar{x}$ solves

$$\begin{aligned} \max:\quad & cx \\ \text{s.t.}\quad & Ax \leq b - e^1 \\ & 0 \leq x \leq d\ , \end{aligned} \tag{9}$$

and $\bar{\bar{x}}$ is a feasible solution of (9). Vector e^1 is defined below.

Assumption: There exists u such that $b_i \geq u > 0$, $\forall i$, and $\varepsilon_1/u < 1$, where

$$\varepsilon_1 = \min\{\varepsilon/2,\ \varepsilon \cdot u/(2M)\}/||c||;\ \text{let } e^1 = (\varepsilon_1, \varepsilon_1, \ldots, \varepsilon_1)^t.$$

Output: Feasible x^* for (7) satisfying $cx^* \geq c\hat{x} - \varepsilon$.

Complexity: $O(||c||^2 \cdot \bar{m} \cdot n \cdot M^4 \cdot N/(\varepsilon \cdot u)^2)$ if $u/M \leq 1$,

$O(||c||^2 \cdot \bar{m} \cdot n \cdot M^2 \cdot N/\varepsilon^2)$ if $u/M > 1$,

where $\bar{m} = \max\{m, \log_2 n\}$ and $N = \max\{\log_2(K/\varepsilon), 1\}$.

Procedure: Apply A2 to (7), using the above ε_1, M, U, K, and $\delta = \varepsilon/2$.

Proof of validity: Define $\lambda = 1 - \varepsilon_1/u$. Since $b_i/u \geq 1$, $\forall i$, and $\varepsilon_1/u < 1$, we have $0 < \lambda < 1$ and $A(\lambda\hat{x}) \leq \lambda b = (1 - \lambda_1/u)b \leq b - e^1$. Trivially $0 \leq \lambda\hat{x} \leq d$, so $\tilde{x} = \lambda\hat{x}$ is feasible for (9). Now $c\hat{x} - c\tilde{x} \leq (1 - \lambda)||c||\ ||\hat{x}|| \leq \varepsilon_1 \cdot ||c|| \cdot M/u \leq \varepsilon/2$ using $M \geq ||\hat{x}||$ and the definitions of λ and ε_1. Hence $c\bar{x} \geq c\tilde{x} \geq c\hat{x} - \varepsilon/2$. Suppose we apply A2 to (7) with the given ε_1, M, U, K, and $\delta = \varepsilon/2$. Since $\tilde{x}$ is feasible for (9), A2 must produce feasible x^* of (7) satisfying $cx^* \geq c\bar{x} - \varepsilon/2 \geq c\hat{x} - \varepsilon$.

The formulas for computational complexity are derived from those for A2 as follows. Parameter δ takes on value $\varepsilon/2$; $\varepsilon_1 = \varepsilon \cdot u/2(2|c| \cdot M)$ if $u/M \le 1$, and $\varepsilon_1 = \varepsilon/(2||c||)$ otherwise. □

Analogous to the remark following A2 one can adapt A3 to the case where some bounds of type $x_j \ge 0$ and/or $x_j \le d_j$ are removed from (7). As before for A2 we then add the assumption "(7) has an optimal solution".

We now apply A3 to the combinatorial problem

$$\begin{aligned} \max: \quad & \bar{1} \cdot x \\ \text{s.t.} \quad & A \cdot x \le \bar{1} \\ & x \ge 0, \quad \text{integral}, \end{aligned} \tag{10}$$

where $m \times n\{0,1\}$ matrix A is the incidence matrix of a normal hypergraph (see Lovász [L12]) and $\bar{1}$ is a vector of all 1's. All basic solutions of the underlying linear program are integral, so we may solve (10) as such. As first step the linear program must be brought into the standard form of (7), so we consider:

$$\begin{aligned} \max: \quad & \bar{1} \cdot x \\ \text{s.t.} \quad & Ax \le b \quad , \\ & x \ge 0 \end{aligned} \tag{11}$$

where $a_i = \tilde{a}_i / ||\tilde{a}_i||$ and $b_i = 1/||a_i||$, $\forall i$. Choose $\varepsilon = 1/2$, $M = n^{\frac{1}{2}}$, $K = U = n$, $u = n^{-\frac{1}{2}}$, and $\varepsilon_1 = n^{-3/2}/4$. With these parameters A3 may be applied to (11), and a feasible x^* of (11) is determined such that $\bar{1} \cdot x^* \ge \bar{1} \cdot \hat{x} - 1/2$, where $\hat{x}$ solves (11). By at most n pivots we obtain a basic (hence integral) x^{**} from $\hat{x}$ by algorithm VERTEX such that $\bar{1} \cdot x^{**} \ge \bar{1} \cdot \hat{x} - 1/2$. (Such pivots require careful implementation if the entire algorithm is to be polynomial--see the method by J.Edmonds [E8].) Hence x^{**} mustbe the desired solution of (11). For the problem at hand $u/M \le 1$ and $\log_2(K/\varepsilon) > 1$, so the computational effort of A3 is of order $O(||c||^2 \cdot m \cdot n \cdot M^4 \cdot \log_2(K/\varepsilon)/(\varepsilon \cdot u)^2)$, i.e., $O(\bar{m} \cdot n^5 \log_2 n)$.

It should be noted that more elaborate versions of A1 can be constructed. For example, one can replace the rather simple step 3 by the analog of step 5. Here we will not include details since as yet we have not been able to reduce the computational complexity by such improvements.

2.5 EQUALITY/INEQUALITY SYSTEMS

Suppose we apply A1 to determine a solution for $Ax \le b$, $0 \le x \le d$. If we want to guarantee that A1 will stop with a solution, then we need an $\varepsilon > 0$ such that $\{x | Ax \le b - \varepsilon,\ 0 \le x \le d\}$ is nonempty. This latter condition implies that the polyhedron $\{x | Ax \le b,\ 0 \le x \le d\}$ has dimension n (provided $d > 0$), and it generally precludes straightforward application of A1 to systems with a mixture of equalities and inequalities. All is not lost, however, if additional information about the structure of the system is known. In that case one may be able to obtain a solution by A1 for a modified system, and a second process my exist that adapts the solution by A1 to the original system.

Let ε_o be a lower bound on the defect of the system $Q = \{x | Ax \le b,\ 0 \le x \le d\}$. Then for any $\varepsilon < \varepsilon_o$, in particular $\varepsilon = 1/2\, \varepsilon_o$, $Q \ne \phi \iff Q_\varepsilon^+ = \{x | Ax \le b + \varepsilon . \bar{1},\ 0 \le x \le d\} \ne \phi$, and there is a many-to-one correspondence between vertices of Q_ε^+ and Q (see [G13], or [A5]). When A is totally unimodular we can take $\varepsilon = 1/2n$, and replace the original system $Ax \le b$, by $Ax \le b + \bar{1} \cdot \varepsilon$.

Let us note that for general linear programs with integar coefficients $\varepsilon = 2^{-L}$ will do (see Chapter 3, and 4).

Algorithm A4.

Input:

1. Linear Program

$$\begin{aligned} \max: \quad & cx \\ \text{s.t.} \quad & Ax = b \\ & 0 \le x \le d \,, \end{aligned} \tag{1}$$

where $m \times n$ matrix A has independent rows and

$$||a_i|| = 1, \ \forall i; \text{ let } c_{max} = \max_j |c_j| > 0.$$

2. Positive $M \ge ||\bar{x}||$, $U \ge c\bar{x}$, $K \ge U - c\bar{\bar{x}}$, where $\bar{x}$ solves (1) and $\bar{\bar{x}}$ is feasible for (1).

3. $\alpha > 0$, $\beta > 0$, $k > 0$ such that

   (i) For every basic infeasible solution $\tilde{x}$ of (1):

$$\tilde{x}_j^B < 0 \quad \text{implies} \quad \tilde{x}_j^B \le -\alpha \,.$$

$$\tilde{x}_j^B > d_j \quad \text{implies} \quad \tilde{x}_j^B \ge d_j + \alpha \,.$$

   (Here x^B denotes the vector of basic x_j of x. Later on x^{NB} will denote the vector of nonbasic x_j. Notation c^B, c^{NB}, and A^{NB} is interpreted similarly.)

   (ii) For every feasible basis B of (1): $|B_{ij}^{-1}| \le \alpha \cdot k$, $\forall i, j$.

   (iii) For every basic feasible solution $\tilde{x}$ of (1): $c\tilde{x} < c\bar{x}$ implies $c\tilde{x} \le c\bar{x} - \beta$.

Assumption: $P = \{x | Ax = b, \ 0 \le x \le d\} \ne \phi$.

Output: Basic optimal x^* for (1).

Complexity: $O(\bar{m}\cdot n\cdot M^2\cdot N/\varepsilon^2)$, where $\bar{m} = \max\{m, \log_2 n\}$, $N = \max\{\log_2(K/\beta), 1\}$ and

$$\varepsilon = \min \begin{cases} \beta/3 \\ \beta/(3\cdot\alpha\cdot c_{max}\cdot k\cdot m^2) \\ 1/(2\cdot m\cdot k) \end{cases} .$$

The formula assumes that the complexity of step 0 below dominates the complexity of steps 1 and 2.

Procedure:

0. Apply algorithm A2 to

$$\begin{aligned} \max: &\quad cx \\ \text{s.t.} &\quad b - e \le Ax \le b + e \\ &\quad 0 \le x \le d \end{aligned} \qquad (2)$$

   with ε, M, U, K as specified above, and $\delta = \beta/3$ $(e = (\varepsilon, \varepsilon, \ldots, \varepsilon)^t)$. The algorithm must stop with a solution, say x^o.

1. Define $b^o = Ax^o$. Starting with x^o, find a basic feasible solution x^1 and corresponding basis B, via VERTEX , for

$$\begin{aligned} \max: &\quad cx \\ \text{s.t.} &\quad Ax = b^o \\ &\quad 0 \le x \le d \end{aligned} \qquad (1')$$

   such that $cx^1 \ge cx^o$.

2. Define x^* by

$$x^{*B} = B^{-1}(b - A^{NB}\, x^{1NB}),$$
$$x^{*NB} = x^{1NB} .$$

Proof of validity: For step 0 we note that (1) corresponds to (8) of A2, so M, U, and K are appropriate and $P_e \neq \phi$ of A2 is equivalent to $P \neq \phi$. Hence A2 must stop with x^o as claimed. Complexity is $O(\bar{m} \cdot n \cdot M^2 \cdot N/\varepsilon^2)$, where $\bar{m}$ and N are as specified above. Validity of step 1 is obvious. Vector x^1 thus obtained is feasible for (2) and $cx^1 \geq c\bar{x} - \delta = c\bar{x} - \beta/3$. Hence $Ax^1 = b^o = b + \varepsilon \cdot f$, where $-1 \leq f_i \leq 1$, $\forall i$. We now turn to step 2, and first prove feasibility of x^*.

By definition of x^1 and x^*:

$$x^{*B} = B^{-1}(b - A^{NB} x^{1NB})$$
$$= B^{-1}(b^o - \varepsilon \cdot f - A^{NB} x^{1NB})$$
$$= x^{1B} - \varepsilon B^{-1} f.$$

Due to condition 3 (ii) and the definition of ε,

$$g_i = \varepsilon || B_{i\cdot}^{-1} f || \leq \varepsilon | B_{i\cdot}^{-1} | | f | \leq \varepsilon \cdot \alpha \cdot k \cdot m < \alpha ,$$

so $|x_i^{*B} - x_i^{1B}| < \alpha$. Now $0 \leq x_i^1 \leq d_i$, $\forall i$, and condition 3(i) imply that $0 \leq x^* \leq d$. Optimality of x^* is proved as follows. By definition of ε and the bound on g_i above we have $g_i \leq \beta/3 \cdot c_{max} \cdot m)$, $\forall i$, so:

$$||cx^* - cx^1|| = ||c^B x^{*B} - c^B x^{1B}||$$
$$\leq ||c^B|| \; ||x^{*B} - x^{1B}||$$
$$= ||c^B|| \; ||g||$$
$$\leq (m^{\frac{1}{2}} \cdot c_{max})(m^{\frac{1}{2}} \cdot \beta/(3 \cdot c_{max} \cdot m))$$
$$= \beta/3.$$

Then $cx^1 \geq c\bar{x} - \beta/3$ implies that $c\bar{x} - cx^* \leq 2\beta/3 < \beta$, and x^* must be optimal due to condition 3(iii). □

Similar to A2 and A3 we may adapt A4 when some bounds of type $x_j \geq 0$ and/or $x_j \leq d_j$ are removed from (1). In that case the assumption "$P \neq \phi$" should be replaced by "(1) has an optimal solution".

For problems with a t.u. coefficient matrix A4 may be specialized as follows. Below we have omitted the upper bound d on x since problems of that type will be encountered later on. One can easily modify A5 for the case where x is explicitly bounded from above.

Algorithm A5.

Input:

1. Linear Program

$$\begin{aligned} \max: \quad & cx \\ \text{s.t.} \quad & Ax = b \\ & x_j \geq 0, \quad j \in J, \end{aligned} \tag{3}$$

where $m \times n$ matrix A has independent rows and is t.u., b and c are integral, and $J \subseteq \{1,2,\ldots,n\}$. Both $b_{max} = \max_i |b_i|$ and $c_{max} = \max_j |c_j|$ are positive and $m \geq 2$.

Assumption: (3) has an optimal solution.

Output: Basic optimal x^* for (3).

Complexity: $O(\bar{m} \cdot m^7 \cdot (n \cdot b_{max} \cdot c_{max})^2 \cdot \log_2(b_{max} \cdot c_{max} \cdot m))$ where $\bar{m} = \max\{m, \log_2 n\}$.

Procedure:

0. Scale rows of A such that (3) becomes

$$\begin{aligned} \max: \quad & cx \\ \text{s.t.} \quad & A^1 x = b^1 \\ & x_j \geq 0, \quad j \in J, \end{aligned} \tag{4}$$

where $||a_i^1|| = 1, \forall i$.

1. Apply algorithm A4 to (4) with the following parameters:

$$M = m^{3/2} \cdot b_{max} ; \quad U = m^2 \cdot b_{max} \cdot c_{max} ;$$

$$K = 2 \cdot U ; \quad \alpha = \beta = 1 ; \quad k = n^{\frac{1}{2}} .$$

<u>Proof of validity</u>: The choice of parameters is justified as follows. By assumption (3) has an optimal solution, hence a basic optimal one, say $\bar{x}$ defined by basis B of A. Then $||B^{-1}_{i\cdot} b|| \le m \cdot b_{max}$ sin B^{-1} is a {0, ± 1} matrix, and $M = m^{3/2} \cdot b_{max} \ge ||\bar{x}||$. Similarly $c\bar{x} \le ||c^B|| \, ||B^{-1} b|| \le$ $\le (m^{1/2} \cdot c_{max}) \cdot (m^{3/2} \cdot b_{max}) = b_{max} \cdot c_{max} \cdot m^2 = U$. Parameter K is also appropriate since $c\tilde{x} \ge - U$ for every basic feasible solution of (3). By the integrality properties of t.u. linear programs we may choose $\alpha = \beta = 1$. So far all parameters are independent of the scaling of A, hence they are valid for both (3) and (4). However the requirement $|B^{-1}_{ij}| \le \alpha \cdot k, \forall i,j,$ where B is a basis of A^1, is influenced by this scaling. Each row of A is divided by $||a_i|| \le n^{\frac{1}{2}}$, hence $|B^{-1}_{ij}| \le n^{\frac{1}{2}}$, and we may choose $k = n^{\frac{1}{2}}$. It is easily confirmed that ε of A4 takes on value $1/(3 \cdot c_{max} \cdot m^2 \cdot n^{\frac{1}{2}})$ with these parameters. The formula for the complexity follows now from A4 by substituting the above expressions.

A lower order of complexity than the one of A5 results when one merely want to find a feasible solution for a t.u. system. We include an algorithm for this case since we will require one later on.

Algorithm A6.

Input:

1. System

$$Ax = b \qquad (5)$$
$$x_j \geq 0,\ j \in J\ ,$$

where $m \times n$ matrix A has independent rows and is t.u., and b is integral. Let $b_{max} = \max_i |b_i|$.

Assumption: (5) has a solution.

Output: Basic feasible x^* for (5).

Complexity: $O(m^6 \cdot n^2 \cdot b_{max}^2)$.

Procedure and proof of its validity: Are easily developed analogously to A4 and A5, and we will omit details. By choosing $M = m^{3/2}\ b_{max}$, $N = 1$, $\varepsilon = (2m\ n^{\frac{1}{2}})^{-1}$, $\bar{m} = m$, and replacing in the complexity formula for A4, we obtain the given formula.

2.6 A FULLY POLYNOMIAL ALGORITHM FOR t.u. LINEAR PROGRAMS

Algorithms A5 and A6 are now combined with a scaling scheme to obtain a polynomial algorithm for

$$\begin{aligned} \max:\ & cx \\ \text{s.t.}\ & Ax = b \\ & x \geq 0 \end{aligned} \qquad (1)$$

where $m \times n$ matrix A is t.u., and $b \geq 0$ and $c \leq 0$ are integral and nonzero. In addition we will assume that (1) has a feasible solution for any nonnegative right-hand side vector. It is not difficult to show that any linear program with t.u. coefficient matrix and integral data may be brought into the form (1) such that all assumptions made about (1) are satisfied (e.g. by addition of artificial variables, bounds on variables, etc.). Furthermore, any such transformation changes problem size and magnitude of data b and c by linear functions in m and n. Hence we can solve any linear program with t.u. coefficient matrix and integral data in polynomial time once we have a polynomial algorithm for (1).

The solution method for (1) proceeds as follows. Algorithm S1 first scales (1) into several linear programs, each of which has a unit vector as right-hand-side. Algorithm S2 takes one of these smaller problems, adds some redundant constraints, and finds a feasible solution for the modified problem. This solution is used to transform the dual of the modified problem into a problem of type (1), so scaling as in S1 becomes possible. This time, however, both objective function and right-hand-side data of each subproblem are well-bounded, and we can solve each such subproblem polynomially. Suppose now the dual of the modified problem has been solved. We then construct a feasible complementary solution for the input problem of S2, which is then passed back to S1. Finally the latter algorithm assembles the outputs of S2 to a solution for (1). Scaling was first employed by J. Edmonds and R.M. Karp in a polynomial algorithm for network flow problem [E 9].

Algorithm S1.

Input: Linear program (1), where $m \times n$ matrix A is t.u., and $b \geq 0$, $c \leq 0$ are integral and nonzero.

Assumption: (1) has a feasible solution for any nonnegative right-hand-side vector.

Output: Optimal integral x^* for (1).

Complexity: $O(d_2(b) \cdot T)$, where $d_2(z)$ is the number of 1's in a binary representation of integral vector z, and T is a bound on the effort of S2 (described below) for solving

$$\begin{aligned} \max:\ & cx \\ \text{s.t.}\ & Ax = u^i \\ & x_j \geq 0,\ j \in J\ , \end{aligned} \tag{2}$$

for any i, $1 \le i \le m$, and $J \subseteq \{1,2,\ldots,n\}$; here u^i is the i'th unit vector.

Procedure:

0. Define D to be a {0,1} matrix of m rows, where row i represents b_i in binary notation (e.g. if $b_i = 7$, then $D_{i.} = (0,0,\ldots,0,1,1,1)$). D has just enough columns so that the largest b_i can be accommodated. Set $k = 1$ and $x^* = 0$.
1. For $i = 1,2,\ldots,m$: If $D_{ik} = 1$, solve (2) with u^i and $J = \{j \mid x_j^* = 0\}$ by (subsequently described) algorithm S2, getting $\hat{x}$; then replace x^* by $x^* + \hat{x}$.
2. Increase k by 1. If D has at least k columns, replace x^* by $2x^*$ and go to 1. Else stop; current x^* solves (1).

Proof of validity: We use double induction on k and i. Since c is nonpositive, $x^* = 0$ is optimal for (1) with zero as right-hand-side. Inducting on k, suppose integral x^* is optimal for (1) with right-hand-side b^{k-1}, which is the decimal interpretation of D^{k-1}; the latter matrix consists of the first k-1 columns of D. (We take $b^{k-1} = 0$ if $k - 1 = 0$.) If the k-th column of D is zero, then $2x^*$ is the optimal solution to (1) with right-hand-side b^k defined from D^k. If $D_{ik} = 1$, inductively suppose that integral x^* is optimal for (1) with right-hand-side $b^{k,i-1}$, the decimal interpretation of $D^{k,i-1}$ obtained from D^k by replacing all 1's in (ℓ,k) position, $\ell = i,i+1,\ldots,m$, by 0. To find the optimal solution for $b^{k,i}$, we now solve

$$\begin{aligned} \max: \quad & c(x^* + x) \\ \text{s.t.} \quad & A(x^* + x) = b^{k,i} \\ & x^* + x \ge 0 \;. \end{aligned} \tag{3}$$

Since $b^{k,i} = b^{k,i-1} + u^i$ and $Ax^* = b^{k,i-1}$, we may equivalently solve

$$\begin{aligned} \max:\quad & cx \\ \text{s.t.}\quad & Ax = u^i \\ & x \geq -x^* \quad . \end{aligned} \tag{4}$$

A relaxed version of (4) is

$$\begin{aligned} \max:\quad & cx \\ \text{s.t.}\quad & Ax = u^i \\ & x_j \geq 0, \quad j \in J \ , \end{aligned} \tag{5}$$

where $J = \{j \mid x_j^* = 0\}$. By induction x^* solves

$$\begin{aligned} \max:\quad & cx \\ \text{s.t.}\quad & Ax = b^{k,i-1} \\ & x \geq 0 \quad , \end{aligned} \tag{6}$$

and we claim it also solves

$$\begin{aligned} \max:\quad & cx \\ \text{s.t.}\quad & Ax = b^{k,i-1} \\ & x_j \geq 0, \quad j \in J \quad . \end{aligned} \tag{7}$$

<u>Proof of claim</u>: Let $\tilde{x}$ solve (7). Since x^* is feasible for (7), we have $cx^* \leq c\tilde{x}$. If $cx^* = c\tilde{x}$, we are done. Otherwise $c(\tilde{x} - x^*) > 0$. Note that $A(\tilde{x} - x^*) = 0$. Since, by definition, for $j \notin J$, $x_j^* > 0$, $\exists\, t_o > 0$ such that $\bar{x}(t) = x^* + t(\tilde{x} - x^*)$ is feasible for (6), for $t \in [0, t_o]$. But, then $c\bar{x}(t_o) = cx^* + t_o c(\tilde{x} - x^*) > cx^*$, contradicting the optimality of x^*. Thus x^* solves (7) too.

We conclude that the dual problem of (7), and hence the one of (6), is feasible. Primal feasibility of (6) follows from assumptions about (1), so (5) has an optimal basic solution $\hat{x}$ (which is furnished by S2. This solution is of the form

$\hat{x}^B = B^{-1} u^i$, $\hat{x}^{NB} = 0$ for some basis B of A. Since B is t.u., $B^{-1} u^i$ is a $\{0, \pm 1\}$ vector, and $\hat{x}$ is also feasible for (4). We conclude that integral $x^* + \hat{x}$ solves (3), which was to be shown. Then by induction S1 does solve (1) with the claimed complexity. □

A glance at the complexity of A5 confirms that A5 may not process (5) in polynomial time, so we must "break down" (5) into even smaller problems. This is done in Algorithm S2.

Algorithm S2.

Input: Linear program (5), where $m \times n$ matrix A is t.u. and $c \leq 0$ is integral and nonzero.

Assumption: (5) has an optimal solution.

Output: Basic optimal x^* for (5).

Complexity: $O(d_2(-c) \cdot n^{10} \cdot \log_2 n)$, where $d_2(\cdot)$ is the function defined in S1.

Procedure:

0. Find a basic feasible solution $\tilde{x}$ to $Ax = u^i$, $x_j \geq 0$, $j \in J$, using A6. Change (5) to the problem

$$\begin{aligned} \min: \quad & -cx \\ \text{s.t.} \quad & Ax = u^i \\ & x_j \geq 0, \quad j \in J \\ & x_j \geq -2, \quad j \notin J \quad . \end{aligned} \tag{8}$$

Since $\tilde{x}$ is a $\{0, \pm 1\}$ solution to (8), the latter problem can be changed to

$$\begin{aligned} \min: \quad & -cx \\ \text{s.t.} \quad & Ax = 0 \\ & x \geq d \ , \end{aligned} \tag{9}$$

where d is the {0,-1,-2,-3} vector defined by

$$d_j = \begin{cases} 0 - \tilde{x}_j & ; \quad j \in J \\ -2 - \tilde{x}_j & ; \quad j \notin J \ . \end{cases} \tag{10}$$

The dual problem of (9) is

$$\begin{aligned} \max: \ & (0|d^t) \begin{bmatrix} y^1 \\ \hline y^2 \end{bmatrix} \\ \text{s.t.} \ & [A^t|I] \begin{bmatrix} y^1 \\ \hline y^2 \end{bmatrix} = -c \\ & y^2 \geq 0 \end{aligned} \tag{11}$$

where the partitions of $y = [y^1|y^2]$ and of the remaining arrays agree.

1. We now solve (11) analogous to the procedure in S1. Define E to be a {0,1} matrix of n rows, where row ℓ represents $-c_\ell$ in binary notation analogous to D and b in S1. Set $k = 1$ and $y^* = [y^{1*}/y^{2*}] = 0$.

2. For $\ell = 1,2,\ldots,n$: If $E_{\ell k} = 1$, solve with algorithm A5

$$\begin{aligned} \max: \ & (0|d^t)y \\ \text{s.t.} \ & [A^t|I]y = u^\ell \\ & y_j^2 \geq 0, \quad j \in L \ , \end{aligned} \tag{12}$$

where $L = \{j|y_j^{2*} = 0\}$, getting $\hat{y}$; replace y^* by $y^* + \hat{y}$.

3. Increase k by 1. If E has at least k columns, replace y^* by $2y^*$ and go to 1.

4. Current y^* solves (11). With algorithm A6 find a basic feasible solution x^* to:

$$
\begin{aligned}
&Ax = u^i \\
&x_j \geq 0 \ , \ j \in J \qquad\qquad (13)\\
&x_j = 0 \ , \ j \ni y_y^{2*} > 0
\end{aligned}
$$

(i.e., drop variables x_j required to be zero from (13) and delete any dependent equations. Apply A6 to the resulting system.) This x^* is a basic optimal solution for (5).

<u>Proof of validity</u>: By assumption, (5) has a basic optimal solution, which must be a $\{0, \pm 1\}$ vector. Hence restriction (8) of (5) admits such a solution as well. Problem (9) is derived from (8) by substituting $(\tilde{x} + x)$ for x, so (9) is equivalent to (8). In particular (9) has a feasible solution. The dual problem (11) of (9) clearly has a feasible solution for any nonnegative right-hand-side, and arguments analogous to those for S1 prove that y^* of step 4 does solve (11). It is easily verified that y^* is also optimal for the dual of (8). Hence y^* and any optimal x^* of (8) are complementary. But there exists a $\{0, \pm 1\}$ x^* for (8), and $x_j^* > -2$, $\forall j \notin J$ implies that the y_j^{2*} corresponding to that inequality must be zero. Suppose we drop all such y_j^{2*}, $j \notin J$ from y^* and delete related inequalities from (8), getting $\bar{y}^*$ and problem (8'), say, where (8') is just a rewrite of (5). Then $\bar{y}^*$ is optimal for the dual of (8'). By complementarity x^* is optimal for (8') iff x^* is feasible for (8') and for $j \in J$, $x_j^* = 0$ whenever $\bar{y}_j^{2*} > 0$. Problem (13) expresses the latter conditions, hence x^* found by A6 for (13) solves (8') and (5).

The complexity calculations are as follows. The effort of A6 in steps 0 and 4 is of order $O(m^6 \cdot n^2)$. Problem (12) in step 2 has n equations and $m + n \leq 2n$ variables. We note that $n \geq \log_2(2n)$, so $\bar{m}$ in the formula of A5 takes on value n, and the effort of A5 is of order $O(n^{10} \log_2 n)$. Now $d_2(-c)$ applications of A5 are required. Combining these terms we get the expression claimed above.

<u>Theorem</u> Complexity of Algorithm S1 with S2 as a subroutine for solving (1) is of order.

$$O(d_2(b) \cdot d_2(-c) \cdot n^{10} \cdot \log_2 n).$$

<u>Proof</u>: Simply combine complexity of S1 with that of S2. □

Finally we remark that a modified version of S1 may be applied to (1) even if A is not known to be t.u. We must then verify that each subroutine produces the expected output, and that the computational effort does not exceed the estimated upper bound (under the assumption that A is t.u.). The resulting version of S1, say S1', is clearly polynomial. It is claimed that either S1' produces an integral optimal solution for (1) or one of the checks proves A to be non-t.u. We will outline a proof. Suppose S1' does terminate with an integral x^*. Feasibility of x^* for (1) follows from the procedure in S1, so we need only show optimality. It is easily verified that x^* is optimal if S2 supplies an optimal solution for each problem of type (5) encountered. But S2 finds feasible complementary solutions for (5) and its dual, so the solution of (5) passed to S1 must indeed be optimal for (5).

Some Remarks

We would like to remark once more that perturbation toward to the inside of the polyhedron, i.e. $x^+ = x' - \varepsilon a_i$, is not essential. One can equally perturb the polyhedron toward the outside, i.e. one can take $x^+ = x'$, the projection of the current point on the chosen hyperplane. Then one must stop when the maximum violation of the current point $\leq \varepsilon$. This is based on the perturbation lemma, which is used in this chapter for totally unimodular linear programs, and developed and discussed in Chapters 3 and 4. What we used in this chapter, by defining $\varepsilon = (3\beta||c||m^2n)^{-1}$, is, in fact, a priori development and a posteriori specialization of the perturbation lemma of Gács-Lovász [G13]. This is also observed recently by Goffin [G38].

One can use Lemma 4.2 in finding a point x s.t. $d(x,K) \leq \varepsilon$, for a convex set K, where $d(x,K) = \inf_{y \in K} d(x,y)$. One has to assume that at every $x \notin K$ there is a subgradient of a function f which represent K as the solution set i.e. $K = \{x: f(x) \leq 0\}$. Equivalently an separation oracle can be used instead of subgradient oracle. By writing the subgradient inequalities at the points of realization of the algorithm, one can apply the above lemma. If $M \geq ||y - x_o||$, where $y \in K$, in at most M^2/ε^2 iteration either one finds an x such that $d(x,K) \leq \varepsilon$, or proves that $K = \phi$. If K is given by a possibly infinite system of inequalities, one has to choose an inequality whose violation $>\varepsilon$, otherwise the current point is an ε-solution, i.e. $d(x,K) \leq \varepsilon$.

In a recent paper, Drezner [D21] observed the following: if at iteration k

$$\sqrt{g_o} > \sqrt{g_k} + ||x_o - x_k|| ,$$

then P is empty, where g_k is defined as in algorithm A0 with $\varepsilon = 0$.

3 The ellipsoidal algorithm

This chapter is in two parts. Part A is devoted to the historical development of the ellipsoidal algorithm. Our aim is to clarify the confusion on what is the ellipsoidal algorithm, and how it was developed and by whom. We give a rather long survey discussing the relaxation methods, cutting plane methods, Levin's algorithm and Mitiagin's lemma, Shor's generalized gradient algorithm, his 1970 and 1977 papers and Judin-Nemirovskii's 1976 papers. We tried to emphasize the fact that the ellipsoidal algorithm is a finite, polynomial time algorithm to find an approximate solution to a general constrained convex minimization problem.

In Part B of the chapter, we discussed Hachiyan's contribution in resolving the question LP $\in P$. We showed that his characterization of nonsolvability of a linear inequality system is an adaptation of a result of Chernikov, and Eremin. We discuss the definition of L, the length of input, and its use in locating the solutions, and in the characterization of nonsolvability. Apart from Chernikov's formula, we give two different proofs of the above characterization. We also discuss the approximate arithmetic aspect of the ellipsoidal algorithm.

We give Hachiyan's proof of the main result, along with our two previously announced proofs of the main result based on works of Shor, and Judin-Nemirovskii [A 5 ,A 7]. We also discuss some well-known proofs of the main result: Gács-Lovász [G 13], Korte-Schrader [K 19], Padberg-Rao [P 3], etc.

PART A: HISTORICAL DEVELOPMENT

3.1 RELAXATION METHODS

The relaxation algorithm of Agmon-Motzkin-Schoenberg [A 3 ,M 27] is the foundation of subgradient optimization in general, and the ellipsoidal algorithm in particular. Suppose we are given a system of linear inequalities in R^n :

$$\ell_i(x) = a_i x + b_i \le 0, \quad i = 1,\dots,m. \tag{1}$$

For simplicity let us assume that we have $||a_i|| = 1$, $\forall i$. Let us define $f(x) = \max_i \ell_i(x)$. Then, an iteration of the relaxation algorithm will be

$$x^+ = x - \lambda \ell_i(x) a_i \ , \tag{2}$$

where $\lambda \in (0,2]$, and i is chosen such that $\ell_i(x) > 0$. Each index selection rule, and rule for choosing λ defines an particular member of the class of relaxation algorithms.

When index i is chosen such that $\ell_i(x) = f(x)$, the algorithm becomes a subgradient algorithm, since the hyperplane represented by $a_i x + b_i \le 0$ is a separating hyperplane for the function $f(x)$, or equivalently a_i is a subgradient of f at x.

Convergence proofs of the relaxation algorithm are based on the following simple geometric lemma, which is due to Agmon.

Lemma 1: Let $x, y \in R^n$, and $\ell_i(x) > 0$, $\ell_i(y) \le 0$. Then for $\lambda \in (0,2]$

$$||x^+ - y||^2 \le ||y - y||^2 - |1-(1 - \lambda^2)| \ell_i^2(x). \qquad \square$$

It is well-known that for $\lambda \in (0,2)$ a relaxation algorithm has a geometric convergence for the sequence $s_k = d(x_k,P)$, whose rate depends on the geometry of the problem through a complex formula. Let

$$P = \{x: Ax \le b\},$$

and

$$d(x,P) = \inf_{y \in P} ||x - y||,$$

then

$$d(x^+,P) \le d(x,P)\theta, \qquad \theta \in (0,1).$$

When the relaxation algorithm is applied to minimize f(x), with $\lambda \in (0,2)$, through formula (2), and with index satisfying $f(x) = \ell_i(x)$, using the fact that f(x) is Lipschitz with constant 1, one obtains that

$$\min_{j \le k} f(x_j) \le K\,\theta^k\, d(x^o,P),$$

provided that $P \ne \phi$. K is some constant.

One must point out two things. First, the sequence $f(x_j)$ is not monotonically nondecreasing in the relaxation and the subgradient algorithms. Second, θ depends on problem parameters through geometry, and $\log\theta$, in general, is not a polynomial function of the length of the input of the problem. This prevents the relaxation algorithm from being a polynomial algorithm. Goffin [G32], gave a thorough discussion of this point.

3.2 CUTTING PLANE ALGORITHMS

The reduction of the feasible region by cutting plane is well-known in mathematical programming. On the one hand Gomory [G45], then many others, used it in order to discard non-integer vertices in integer programming. The other direction is in convex programming. The basic idea is to replace a convex function locally by one of its separating hyperplanes. By using objective function cuts, one obtains a sequence of problems, with decreasing feasible sets, and increasing objective function value. The method is developed from the Chebyshev curve fitting problem, and was developed simultaneously by Kelley [K13], and Cheney and Goldstein [C15].

It is known that the method converges, though its convergence is slow. Elzinga and Moore [E11] showed that under suitable assumptions the subsequence of feasible points has a geometric convergence, see also Eaves and Zangwill [E1], Luenberger [L18].

3.3 LEVIN'S ALGORITHM AND MITIAGIN'S LEMMA

A.Ju. Levin [L11] considered the approximate minimization of a convex function over a convex polytope M_o. He called his algorithm the method of central sections. He assumes f is differentiable, since a convex function is differentiable almost everywhere. Clearly f is Lipschitz on M_o. If $\nabla f(x_o) = 0$, then x_o is a global minimizer of f. Otherwise we know that x^*, a minimizer of f, lies in $\{x: \langle \nabla f(x_o), x - x_o \rangle \le 0\}$. So we can take the new polytope as $M_1 = \{x \in M_o: \langle \nabla f(x_o), x - x_o \rangle \le 0\}$ Let x_1 be the centroid of M_1. Check $\nabla f(x_1)$ and continue.

The basic idea is to make the next polytope as small as possible. The best point to consider is the centroid of the polytope. Reasoning comes from the following theorem of B.S. Mitiagin [M26].

Theorem 1: Let S be a convex body in R^n, $\bar{x}$ its centroid; $a \in R^n$, $a \neq 0$, and

$$S_1 = \{x \in S : \langle a, x - \bar{x} \rangle \leq 0\}, \quad S_2 = \{x \in S : \langle a, x - \bar{x} \rangle \geq 0\},$$

then

$$\frac{\mathrm{vol}(S_1)}{\mathrm{vol}(S_2)} \geq \left(\frac{n}{n+1}\right)^n, \tag{1}$$

and consequently,

$$\frac{\mathrm{vol}(S_i)}{\mathrm{vol}(S)} \leq 1 - \left(\frac{n}{n+1}\right)^n \leq 1 - e^{-1}, \quad i = 1,2 \tag{2}$$

□

When S is a cone of revolution, the bound in (1) is exact, and the proof is trivial. For a general convex body, Mitiagin uses a symmetrization due to Brunn-Minkowskii, or Steiner (see [G52], [E 10]).

For $n = 1$ the above method reduces to binary search (by halfing the interval), and in $\log (b-a)K/\varepsilon$ iterations one obtains an ε-optimal solution, i.e. $f(\bar{x}) - f(x^*) \leq \varepsilon$, where x^* solves f, $S = [a,b]$, and K is the Lipschitz constant for f.

For general n, $q = 1 - e^{-1}$ is a guaranteed rate of reduction of successive polytopes independent of n. Thus in k iterations one reduces the volume of the current polytope to $v_o q^k$, where v_o volume of the initial polytope. Consequently we have the following estimate:

$$\min_{\ell \leq k} f(x_\ell) - f(x^*) \leq K\left(\frac{v_o}{\omega_n}\right)^{1/n} q^{k/n},$$

where $\omega_n = \Pi^{n/2}\, \Gamma(\frac{n}{2} + 1)$ is the volume of the unit ball in R^n.

It is known that a convex polytope can be imbedded in a simplex such that the volume of the simplex is less than equal γ_n times the volume of the polytope. Here γ_n is an absolute constant depending on the dimension n. Levin suggested the alternative procedure; in every q steps replace the given polytope with a simplex with the above requirements. Obviously it is easier to work with simplices than polytopes. Recently Yammitsky [Y 2], and Yammitsky-Levin [Y 3] (a different Levin) showed that one can work only with simplices. The price one pays is the relatively small volume reduction ratio. Remarkably, this ratio is roughly $\exp\{1/2(n+1)^2\}$, thus allowing a polynomial algorithm for approximate minimization of convex functions. This new algorithm will be studied in Chapter 8. Note also that Newman [N6] suggested a similar algorithm.

3.4 SHOR'S GENERALIZED GRADIENT METHOD

Shor in [S 9] introduced the generalized gradient method. Today we call it the subgradient method. Let f(x) be a convex function. We do not assume differentiability. In fact we study the case when f is non-differentiable.

Let us first introduce subgradients. For $f: R^n \to R$, $u \in R^n$ is said to be a subgradient of f at x, i.e. $u \in \partial f(x)$ if

$$f(y) - f(x) \geq \langle u, y - x \rangle, \quad \forall y \in R^n .$$

When f is differentiable $\partial f(x) = \nabla f(x)$, the gradient. The generalized gradient method is a generalization of the steepest descent method i.e.

$$x_{k+1} = x_k - \gamma_k \nabla f(x_k) ,$$

and γ_k is chosen to minimize f on the line generated by the gradient. In other words, $\gamma_k = \underset{\gamma}{\operatorname{argmin}}\, f(x_k - \gamma \nabla f(x_k))$. Under mild conditions for a general f, x_k converges to x^*, a local minimum of f (when f is convex to a global minimum), and its rate of convergence is linear and depends on the condition number of the Hessian of f near x^*.

A subgradient can be described geometrically as the normal of the supporting hyperplane of the level curve of the function.

A basic difference between the generalized gradient and the steepest descent is that in the steepest descent function values decrease in succesive iterations, whereas in generalized gradient this is not true. As a result of this the choice of step length is quite different in both methods. A typical iteration of the generalized gradient is as follows:

$$x_{k+1} = x_k - \gamma_k \frac{u_k}{||u_k||} \quad .$$

The usual requirements on γ_k to ensure convergence or near-convergence of x_k or $f(x_k)$ to x^* or $f(x^*)$ are

$$\lim_{n\to\infty} \sum_{k=1}^{n} \gamma_k = \infty, \quad \gamma_k > 0, \quad \lim_{k\to\infty} \gamma_k = 0$$

(see [P12, E24]).

For a convex function Shor [S11] defined a condition number σ as

$$\sup \frac{||x - x^*||}{||y - x^*||} \leq \sigma$$

for $f(x) = f(y) \neq f(x^*)$; x^* is a minimizer of f.

Then Shor proved the following:

Theorem 1: Let $||x^* - x_o|| \leq h\,\sigma$, $\sigma \geq \sqrt{2}$

$$x_{k+1} = x_k - h_k\, g_k / ||g_k||, \quad g_k \in \partial f(x_k)$$

$$h_{k+1} = h_k \sqrt{\frac{\sigma^2 - 1}{\sigma^2}} \quad .$$

Then x_k is a convergent sequence to x^* and $||x_k - x^*|| \leq h_k \sigma$.

□

Remark: Since any convex function is Lipschitz on bounded sets, by letting K be a Lipschitz constant of f on $\{x : ||x - x_o|| \leq h_o\}$ we have

$$f(x_k) - f(x^*) \leq K\, ||x_k - x^*|| \leq K h_o\, \sigma^k \; .$$

Hence $f(x_k) - f(x^*)$, in particular $\min_{\ell \leq k} f(x_\ell) - f(x^*)$, has a geometric convergence rate, however slow. Moreover the convergence rate and the condition number depends on the function.

3.5 SHOR'S SPACE DILATATION OPERATOR

N.Z. Shor, in two papers in 1970 ([S13, S14]), introduced the space dilatation operator and associated algorithm. More precisely, in Soviet literature the method is introduced in 1969 (see [S12]). The purpose was to increase the rate of convergence of generalized gradient, especially for gully type functions. Gully functions are those for which a generalized gradient descent method does not have geometric rate of convergence because the condition number σ may be infinite.

The method can be seen as a generalized gradient method with a variable metric (see Bland-Goldfarb-Todd [B24], [G 37]. At each iteration we associate a different argument space via a space dilatation coefficient which, in general, varies from iteration to iteration.

Let us describe these mathematically.

Preliminaries

Let $f: R^n \to R$ be a convex function. A vector $u \subset R^n$ is said to be a subgradient at x if

$$f(y) - f(x) \geq \langle u, y-x \rangle, \quad \forall y \in R^n, \tag{1}$$

or equivalently

$$f(x) \leq f(y) + \langle u, x-y \rangle . \tag{2}$$

The set of subgradients of f at x will be denoted by $\partial f(x)$, i.e.

$$\partial f(x) = \{u \in R^n : f(y) - f(x) \geq \langle u, y-x \rangle, \quad \forall y \in R^n\} .$$

Here $\langle \cdot, \cdot \rangle$ is the usual inner product. By $g_f(x)$ we will denote an arbitrary element of $\partial f(x)$; g will always mean a subgradient vector.

Let T: $R^n \to R^n$ be a non-singular linear transformation. We will denote by y, y = Tx, elements of the new space i.e. transformed space. Let us define Ψ as

$$\Psi(y) = f(T^{-1} y) \quad , \tag{3}$$

i.e.

if $T(x) = y$ then $f(x) = \Psi(y)$

Lemma 1: Let f: $R^n \to R$ convex, T: $R^n \to R^n$ non-singular. Then

(i) $\Psi(y)$ is convex ,

(ii) $g_\Psi(y) = (T^{-1})^t \, g_f(x)$ or equivalently

$$\partial\Psi(y) = (T^{-1})^t \partial f(x) \quad . \tag{4}$$

Proof:

(i) For $\alpha + \beta = 1, \quad \alpha \geq 0, \quad \beta \geq 0, \quad y = Tx_1, \quad y_2 = Tx_2$

$$\begin{aligned} \Psi(\alpha y_1 + \beta y_2) &= f(T^{-1}(\alpha y_1 + \beta y_2)) = f(\alpha x_1 + \beta x_2) \\ &\leq \alpha f(x_1) + \beta f(x_2) = \alpha f(T^{-1} y_1) + \beta f(T^{-1} y_2) \\ &= \alpha\Psi(y_1) + \beta\Psi(y_2) \quad . \end{aligned}$$

(ii) Let $y_0 = Tx_0$, $y = Tx$; then

$$\begin{aligned} \Psi(y) - \Psi(y_0) &= f(x) - f(x_0) \geq \langle g_f(x_0), \; x - x_0 \rangle \\ &= \langle g_f(x_0), T^{-1}(y - y_0) \rangle = \\ &\qquad \langle (T^{-1})^t \, g_f(x_0), \; y - x_0 \rangle \quad , \end{aligned}$$

i.e. we have

$$\Psi(y) - \Psi(y_0) \geq \langle (T^{-1})^t \, g_f(x_0), y - y_0 \rangle,$$

which implies

$$(T^{-1})^t \; g_f(x_0) \in \partial\Psi(y_0) \; .$$

Let $g_\Psi(y_0) \in \partial\Psi\,(y_0)$; then

$$\Psi(y) - \Psi(y_0) \geq \langle g_\Psi(y_0), \; y - y_0 \rangle$$

$$f(x) - f(x_0) \geq \langle g_\Psi(y_0), \; Tx - Tx_0 \rangle = \langle T^t g_\Psi(y_0), \; x - x_0 \rangle \; ,$$

which implies

$$T^t \; g_\Psi(y_0) \in \partial f(x_0) \; ,$$

i.e. $T^t \partial\Psi \; \partial f \Rightarrow \partial\Psi \subset (T^{-1})^t \; f$.

$\therefore \quad \partial\Psi = (T^{-1})^t \partial f$.

Since the subgradient of a function at any point is not unique, (4) must be understood in the sense that there is a 1-1 correspondence between elements of two sets.

Let $s \in R^n$, $\langle s,s \rangle = 1$; then the <u>space dilatation operator</u> in the direction s is

$$R_\alpha(s) = I + (\alpha - 1)ss^t \; . \tag{5}$$

Let $X = \langle s \rangle = \{\alpha s \,|\, \alpha \in R\}$. Then $R^n = X \oplus X^\perp$. Then $x \in R^n$ can be uniquely written as

$$x = x_s + \bar{x}_s \; , \quad x_s \in X, \quad \bar{x}_s \in X^\perp \; .$$

Then

$$R_\alpha(s)(x) = \alpha x_s + \bar{x}_s \; ,$$

i.e. $R_\alpha(s)$ multiplies the s-component x_s of x by α , leaving the orthogonal complement fixed.

Lemma 2:

(i) $R_{\alpha\beta}(s) = R_\alpha(s) R_\beta(s)$, (6)

(ii) $R_\alpha(s) R_{\frac{1}{\alpha}}(s) = R_1(s) = I, \quad \alpha \neq 0$,

(iii) $||R_\alpha(s)x||^2 = ||x||^2 + (\alpha^2 - 1) < x,s >^2$. (7)

Proof:

(i) can be verified by direct computation. (ii) follows from (i) and the definition.

(iii) $||R_\alpha(s)x||^2 = < x + (\alpha - 1)(s,x), \quad x + (\alpha - 1)< s,x > s >$

$= < x,x > + 2(\alpha - 1)< s,x > < s,x > + (\alpha - 1)^2 < s,x >^2 (s,s)$

$= ||x||^2 + < s,x >^2 \{(\alpha - 1)^2 + 2(\alpha - 1)\} =$

$= ||x||^2 + (\alpha^2 - 1)< s,x >^2$.

Lemma 3: $\det R_\alpha(s) = \alpha$.

Proof: Since for $x \in X$, $R_\alpha(s)x = \alpha x$, and for $x \in X^\perp R_\alpha(s)x = x$, eigenvalues of $R_\alpha(s)$ are α , and 1. The eigenvalue 1 has multiplicity (n-1). Thus

$$\det R_\alpha(s) = \alpha \cdot 1 \cdot 1 \cdot \cdot \cdot 1 = \alpha.$$

The algorithm GGSD

(Generalized Gradient with Space Dilatation)

0. $x_o \in R^n$, $B_o = I$, $k = 0$.

1. Let $g(x_k) = g_f(x_k) \in \partial f(x_k)$;

if $g(x_k) = 0$, stop; x_k is a minimizer.

2. Let $s_k = B_k^t \, g(x_k)/||B_k^t \, g(x_k)||$. (8)

3. $x_{k+1} = x_k - h_k B_k s_k$. (9)

4. $B_{k+1} = B_k R_{\beta_k}(s_k)$. (10)

5. $k = k + 1$, go to 1.

Where $\beta_k = \alpha_k^{-1}$, α_k is the space dilatation operator, $\alpha_k > 1$. At each iteration α_k, h_k are determined.

Let us note for $H_k = B_k B_k^t$, we can write basic iterations as

$$x_{k+1} = x_k - h_k \frac{H_k g_k}{\sqrt{g_k^t H_k g_k}} = x_k - h_k \frac{H_k g_k}{||g_k||_k} , \tag{9'}$$

$$H_{k+1} = H_k + (1 - \beta_k^2) \frac{(H_k g_k)(H_k g_k)^t}{g_k^t H_k g_k} \tag{11}$$

$$= H_k + (1 - \beta_k^2) \frac{(H_k g_k)(H_k g_k)^t}{||g_k||_k^2} , \tag{11'}$$

where $||\cdot||_k$ is the norm induced by the metric defined by H_k; the above symmetric form representation was first noticed by Skokov [S 36].

Let us remark that the above description covers a family of algorithms. Each choice of α_k, h_k will give a specified algorithm. In particular, by suitable choices of α_k, h_k, one obtains the ellipsoidal algorithm. In this sense, the ellipsoidal algorithm is an instance of GGSD.

Let $M^* = \{x: f(x) = \min_{x \in R^n} f(x)\}$. Shor proved the following:

Theorem 4: Let us assume M^* is bounded, $\alpha_k \geq 1 + \varepsilon$, $\varepsilon > 0$, and $\inf_{x^* \in M^*} ||A_k(x_k - x)|| \leq r$ (12)

for some positive r. Then

$$\lim_{k \to \infty} f(x_k) = f(x^*), \quad x^* \in M^*. \qquad \square$$

Let us introduce the notation

$$\rho(x) = \min_{x^* \in M^*} ||x - x^*|| = \min_{x^* \in M^*} ||x - x^*(x)|| .$$

Let m^* be such that

$$f(x^*) = m^* , \text{ for } x^* \in M^*, \text{ and}$$

$f'_s(x) = \sup\{< s,u > , \ u \in \partial f(x)\}$, the directional derivative of f in the direction s, and

$$x^{**} = \operatorname{argmin} \{ ||x^* - y|| : y = x + t(x - x^*), \ t \geq 0\}.$$

The next theorem gives sufficient conditions for Theorem 4 to hold

Theorem 5: Let M be bounded, and for all x such that $\rho(x) \leq \rho(x_o) = R$, and $x^* \in M^*$, the following is satisfied:

$$N[f(x)-m^*] \leq |f'_{x-x^*}(x)| \ ||x - x^{**}|| \leq M[f(x)-m^*]. \qquad (13)$$

Then, if in applying GGSD we have:

(i) $\rho(x_o) \leq R$,

(ii) $h_{k+1} = \frac{2MN}{M+N} \frac{[f(x_k)-m^*]}{||B_k^t g_f(x_k)||}$,

(iii) $1 + \varepsilon \leq \alpha_{k+1} \leq \frac{M+N}{M-N}$,

then the hypothesis of Theorem 4 will be satisfied (here M, N, ε ,R are positive constants). □

When f has a unique minimum at x^* , we obtain the following as a corollary to the above theorem.

Theorem 6: Let f have a unique minimizer, say x^*, and $f(x^*) = m^*$, and for all x satisfying $\rho(x) \leq R$, the following is satisfied

$$N[f(x)-m^*)] \leq |f'_{x-x^*}(x)| \; ||x - x^{**}|| \leq M[f(x)-m^*]. \qquad (14)$$

Furthermore if during the algorithm we have:

(i) $\rho(x_o) \leq R,$

(ii) $h_{k+1} = \frac{2MN}{M+N} \frac{[f(x_k)-m^*]}{||B_k^t g_k(x_k)||}$,

(iii) $1 + \varepsilon \leq \alpha_{k+1} \leq \frac{M+N}{M-N}$,

then $\lim_{x\to\infty} f(x_k) = m^*$. □

Let us remark that the above version of GGSD algorithm depends on the function f(x). First we need to know $m^* = \min_{x \in R^n} f(x)$. Secondly we need the constants M, N. It should be noted that M, N are not dependent on the "gullying coefficient" or the condition number σ , and are invariant under linear transformation.

It is stated that the volume of region of location of optimal solutions decreases as

$$V_k = V_o \prod_{i=1}^{k} \beta_i,$$

and for

$$\alpha_k = \alpha = \frac{M+N}{M-N}$$

it has a geometrical progression with constant $\beta = \alpha^{-1} < 1$. As we will see later in this chapter, in the case of ellipsoidal algorithm volume reduction ratio $\sigma^n \beta$, where here σ

is a different meaning, namely magnification coefficient.

When $f(x) = x^t Kx$, where K is a positive definite matrix, $M = N = 2$, and α_k can be taken arbitrarily large. If $\alpha = \infty$, then the method becomes a conjugate gradient method and minimizes f in n-steps. This is also observed and made explicit by Skokov [S36].

In the second paper Shor [S14] investigated the behaviour of the algorithm (GGSD) when m^* is not known in advance.

<u>Theorem 7</u>: For a convex f, let us assume the following

(i) if $f_1(\alpha) = f[(1 - \alpha)x_1 + \alpha x_2)]: [0,1] \to R$ is strictly decreasing, then

$$f'_{x_1 - x_2}(x_1) \,||x_1 - x_2|| \le M\, [f(x_1) - f(x_2)],$$

for some $M > 1$,

(ii) during the course of GGSD α , h are chosen as

$$\alpha_{k+1} = \frac{M+1}{M-1}$$

$$h_{k+1} = \frac{2M[f(x_k) - m]}{(M+1)\,||B_k^t g_f(x_k)||}\,.$$

Then the following holds:

(a) $m \ge m^*$ implies that $\{h_k\}$ is bounded, and for $\varepsilon > 0$, $\exists\, \bar{k}$ s.t. $f(x_k) < m + \varepsilon$,

(b) $m < m^*$ implies that $\{h_k\}$ is unbounded. □

The above theorem gives an algorithm for determining m^*. Starting with x_o, $m_o = f(x_o)$, $\Delta > 0$, $H > 0$ (large), $r = 0$, for $m = m_k$, either we find eventually $\min_k f(x) \le m_r - \Delta/2$, or $h_k > \Delta$. Depending on the outcome we update, x_r, m_r, Δ_r. It can be shown that $\lim_k \inf_{\ell \le k} f(x_k) = m^*$.

Shor also proved the following theorem which states that there exists a subsequence with a geometric convergence rate.

Theorem 8: Suppose in the execution of the GGSD we have

(i) $\alpha_k = \alpha > 1, \forall k$,

(ii) $\inf_{\{x^*:f(x^*)=m^*\}} ||A_k(x_k - x^*)|| \le r$: for some $r > 0$.

Then there exists an infinite sequence $\{x_{k_i}\}$, $k_i < k_{i+1}$, $\forall i$, such that $f(x_i)-m^* \le c\, \alpha^{-k_i/n}$.

It is observed that GGSD has a geometric convergence rate $q = \alpha^{-1/n}$, independent of the properties of f(x), and can easily be applied, especially for solving systems of nonlinear equations and inequalities.

Judin and Nemirovskii [J 8] referred to this second paper in their study of the ellipsoidal algorithm. Following them, Hachiyan referred to the same paper. A lot of researchers referred to this paper as a source of GGSD algorithm. The fact is, as shown above, the [S13] paper is the main source of the method.

3.6 JUDIN-NEMIROVSKII 1976 PAPERS

D.B. Judin and A.S. Nemirovskii published in 1976 two seminal papers, part of 5 papers [J 7, J 8, J 9, J10, N 2] (see also [N 3]), in which they study the computational complexity of continuous mathematical programming problems. Studies of a similar nature can be found in Ivanov [I3] and sources given in it, and Kuzevkin and Tihomirov [K32]. The papers are of a general nature and consider classes of functions and algorithms. The main objective of the papers is to give upper and lower complexity estimates for the classes of problems.

The ellipsoidal algorithm is a part of these 1976 papers.

Let us consider the following mathematical programming problem:

$$P_o \quad \min_{x \in G} f_o(x) \quad \text{s.t.} \quad f_i(x) \le 0, \quad i = 1,\dots,m.$$

For our purposes we may assume G is compact and convex, and f_i, is convex and defined at least on G, $i = 1,\dots,m$.

Let B^M be an algorithm to solve P_o in (at most) M steps. Let $X^*(B^M, \vec{f})$ be the set of results of B^M for P_o. Note $\vec{f} = (f_o, f_1, \dots, f_m)$. $* \in X^*(B^M, \vec{f})$ denotes P_o is infeasible, i.e. $\nexists\, x \in G$ such that $f_1(x) \le 0$, $i = 1,\dots,m$. Let us define $\varepsilon(B^M, f)$, the defect of B^M on $\vec{f}$ as

$$\varepsilon(B^M, \vec{f}) = \begin{cases} +\infty, & \text{if } \vec{f} \text{ is compatible, and } B^M \text{ does not recognize it or } \vec{f} \text{ is incompatible but } X^*(B^M, \vec{f}) \ne \{*\} \\ \sup\limits_{x \in X^*(B^M, \vec{f})} f_o(x) - f* & \text{otherwise} \end{cases}$$

where f* is the value of an optimal solution.
Similarly we denote the *-defect of B^M on $\vec{f}$ as

$$\varepsilon^*(B^M, \vec{f}) = \begin{cases} +\infty & \vec{f} \text{ comp., } * \in X^*(B^M, \vec{f}) \\ 0 & \vec{f} \text{ incomp., } X^*(B^M, \vec{f}) = \{*\} \\ \sup\limits_{x \in X^*(B^M, \vec{f}) \cap G} \max\{f_o(x) - f*,\ f_i(x),\ i = 1,\dots,m\} & \text{otherwise.} \end{cases}$$

Let U be a class of problems. Let

$$\varepsilon(B^M, U) = \sup_{\vec{f} \in U} \varepsilon(B^M, \vec{f}), \qquad \varepsilon^*(B^M, U) = \sup_{\vec{f} \in U} \varepsilon^*(B^M, \vec{f}).$$

Let us define relative defect and relative *-defect as

$$\omega(B^M,U) = \frac{\varepsilon(B^M,U)}{r(U)}, \quad \omega^*(B^M,U) = \frac{\varepsilon^*(B^M,U)}{r(U)},$$

where $r(U) > 0$ is the absolute $*$-defect of the trivial method with $m = 0$. Let us note that by the above equations there is one to one correspondence between ε and ω .

Let us define '<u>informational complexity</u>' of a class of problems U for a given information transformation or a family of algorithms as

$$M(\varepsilon) = 1 + \sup\{M | \forall B^M : \varepsilon(B^M,U) > \varepsilon\} . \tag{1}$$

In other words, the ε-complexity of a class U for algorithms B^M is the smallest M for which there exists an M-step method to solve problems of class U, such that each problem in that class has an error less than or equal to ε .

Similarly one can define $N(\omega)$ as

$$N(\omega) = 1 + \sup\{M | \forall B^M : \omega(B^M,U) > \omega\} . \tag{2}$$

One can define $M^*(\varepsilon)$, $N^*(\varepsilon)$ similarly. We will deal mainly with $N^*(\omega)$.

<u>Definition</u>: The variation of f_i on G

$$V_G f_i \equiv \sup_{x \in G} f_i(x) - \inf_{x \in G} f_i(x) .$$

Let us consider the constrained convex minimization problem:

$$\min_{x \in G} \; f_o(x)$$
$$\text{s.t.} \quad f_1(x) \le 0,$$

where $G \subseteq R^n$. Let V_i be an upper bound for $V_G f_i$, for $i = 1,2$. This problem will be represented compactly as

$$\vec{f} \in C(G, R^n, 1, V_o, V_1).$$

Here 1 in the third element of the paranthesis represents the number of constraints.

Let us remark that the above formalism is general enough to include P_o. Letting

$$s(x) = \max_{1 \le i \le m} f_i(x) \quad , \tag{3}$$

P_o reduces to

$$\min_{x \; G} f_o(x)$$

$$\text{s.t.} \quad s(x) \le 0.$$

By renaming s as f_1 we obtain the desired result.

The following lemma is essential, and is due to Judin-Nemirovskii.

Lemma 1: Let $G \subseteq R^n$, H be an open set in G, $f \in C(G, R^n, 1, V_o, V_1)$ and $\omega > 0$. Let $a \le \infty$ be such that

$$\text{(i)} \quad \forall x \in G \setminus H, \; f_1(x) \le \omega V_1 \implies f_o(x) \ge a \tag{4}$$

(ii) $\left(\frac{|H|}{|G|}\right)^{1/n} \le \omega$, $H \ne G$. ($|G|$ denotes the volume of G)

Then the optimal value f* of the objective function of problem $\vec{f}$ gives an estimate of $f^* \ge a - \omega V_o$.

Proof: Let x_o be a solution.

For $x_o \notin H$, by (i) $f^* = f_o(x_o) \ge a \ge a - \omega V_o$.

For $x_o \in H$, let $S = \{x: ||x|| = 1\}$. Let dS(e) represent the (n-1) dimensional Lebesque measure on S. For $e \in S$, let

$$\rho_G(e) = \sup\{t | x_o + te \in G\},$$

$$\rho_H(e) = \sup\{\tau \mid x_o + \tau e \in H, \quad 0 \le \tau \le t\}.$$

Let us recall that the volume of G in R^n can be computed by polar coordinates as (see [G52])

$$|G| = \frac{1}{n} \int_S \rho_G^n(e)\,ds(e),$$

taking the coordinate system centered at x_o. Similarly $|H| \ge \frac{1}{n} \int_S \rho_H^n(e)\,dS(e)$; (5) implies that

$$\int_S \rho_H^n(e)\,dS \le \omega^n \int_S \rho_G^n(e)\,dS. \qquad (6)$$

Since $\rho_H(e) \le \rho_G(e)$ for $e \in S$, we can restrict the above integral to $\tilde{S} = \{e \in S\colon \rho_G(e) > 0\}$. Then we must have $\tilde{e} \in \tilde{S}$ such that

$$\rho_H^n(\tilde{e}) \le \omega^n \rho_G^n(\tilde{e}),$$

since otherwise we obtain a contradiction to (6).

Let $y = x_o + \rho_H(\tilde{e})\tilde{e}$, $z = x_o + \rho_G(\tilde{e})\tilde{e}$. Since $\omega < 1$, $y \ne z$, and since H is open in G, $y \notin H$.

Let f be a continuous convex function on G. Considering it on the interval $[x_o, z]$ we get:

$$f(y)-f(x_o) \le \frac{||y - x_o||}{||z - x_o||}(f(z)-f(x_o)) \le \omega V_{G_f}. \qquad (7)$$

Setting $f = f_1$, we find that

$$f_1(y) \le \omega V_1.$$

By (4) $f_o(y) \ge a$.

Taking this inequality for $f = f_o$, and considering $f_o(y) \geq a$ we have

$$f_o(y) - f_o(x_o) \leq \omega V_o ,$$

$$f* = f_o(x_o) \geq f_o(y) - \omega V_o \geq a - \omega V_o. \qquad \square$$

Next we will present a result on the Method of Center of Gravity (MCG) or Levin's algorithm. It is a typical result in the Judin-Nemirovskii papers [J 7, J 8]. Moreover it is a central theorem in the complexity of convex programming in finite dimension.

<u>Theorem 2</u>: The complexity of the class $C(G, R^n, 1, V_o, V_1)$ satisfies, for $\omega > 0$:

(i) $N(\omega) = \infty$

(ii) $N^*(\omega) \leq C_o n \, \ell n \frac{1}{\omega}$, $C_o = \ell n \frac{e}{e-1} \leq 2.2$

(iii) If G is a parallelopided, then

$$N^*(\omega) \geq \frac{n}{3\ell n2} \ell n \frac{1}{14\omega} - 1. \qquad \square$$

Let us define asphericity of G as

$$\alpha(G) = \inf \{\alpha | \exists (x_o, r) : \{||y - x_o|| \leq r\} \subset G$$
$$\subset \{||y - x_o|| \leq \alpha r\}\}.$$

Then $\alpha_2(G)$ is the exact lower bound for asphericity of images of G under all linear isomorphism. By a well-known result of Fritz John, $\alpha_2(G) \leq n$, [J 3] .

Corollary 3: Let $G \subseteq R^n$, $\alpha_2(G) \leq \alpha$, then for the class $C(G,R^n,1,V_o,V_1)$, we have

$$N^*(\omega) \geq \frac{n}{3\ln 2} \ln \frac{1}{28\,\omega\alpha\sqrt{}} - 1 . \qquad \square$$

The theorem states that there is no finite algorithm for $C(G,R^n,1,V_o,V_1)$ which will give an approximate solution with the specified error, which is also feasible. It shows that perturbation of the constraints is necessary. Furthermore it also shows that in this class there is no finite algorithm to solve the unconstrained convex minimization problem exactly. The proof is based on a rather complicated lemma which says that, say for Levin's algorithm, given a finite number of points which realizes such an algorithm, then there exists a convex function positive on the generated points, and zero in some subset of the last region. For ε-optimal solutions or solutions with ε^*-defect, a lower bound on complexity is $O(n \ln \frac{1}{\omega})$. On the other hand MCG (Levin's algorithm) is optimal in the number of iterations among the algorithms which only use $f_i(x)$, $g_i(x) \in \partial f_i(x)$, $i = 0,1$.

The ellipsoidal algorithm

Let us note that the problem of finding the center of gravity of a polytope is almost as difficult as the original optimization problem. Thus it is very desirable from a computational and practical point of view to have implementable versions of the above method (MCG). Judin-Nemirovskii's answer is the ellipsoidal algorithm. They (also) called it the Modified Method of Center of Gravity (MMCG). It roughly replaces each step of MCG with a n-step algorithm retaining the convergence properties of the former.

The following is crucial in the algorithm:

Lemma 4: Let W_r be a ball of radius r in R^n with center at 0. Let $e_1, e_2, \ldots, e_n$ be an orthonormal basis. Let

$$Q(\Phi) = \{x \mid \langle x, e_n \rangle \geq \operatorname{Cos}\Phi \|x\|\}, \quad 0 \leq \Phi \leq \frac{\pi}{2}.$$

Then for $\cos \Phi < \frac{1}{n}$, the set $W_r \setminus Q(\Phi)$ may be included in an ellipsoid of volume $\beta^n(\Phi)|W_r|$ having center at the point $(-r\,\gamma(\Phi)e_n)$, where

$$\gamma(\Phi) = \frac{1-n \operatorname{Cos}\Phi}{n+1},$$

$$\beta(\Phi) = 2(\operatorname{Sin}\frac{\Phi}{2})^{n-1/n} (\operatorname{Cos}\frac{\Phi}{2})^{(n+1)/2} \sqrt{\frac{n^2}{n^2-1}} (\frac{n-1}{n+1})^{1/2n}.$$

When $\Phi = \frac{\pi}{2}$, $Q(\Phi)$ is a half sphere, and we have the formulae

$$\gamma = \frac{1}{n+1},$$

$$\beta = \sqrt{\frac{n^2}{n^2-1}} (\frac{n-1}{n+1})^{1/2n}.$$

Let us define $f_2(x) = \rho(x,G)$, the distance function to G, for $x \notin G$. Clearly, for $x \in G, \rho(x,G) = 0$.

We will assume that G is an ellipsoid.

Let $\omega < 1$ be the required precision. We will choose the coordinates O_1 so that W_1 w.r.t. O_1 will be a unit sphere with center 0. Let $r_1 = 1$. At x_i, let

$$j(i) = \begin{cases} 2 & \text{if} \quad x_i \notin \operatorname{int}(G) \\ 1 & \text{if} \quad f_1(x_i) > \omega V_1, \text{ and } x_i \in \operatorname{int}(G) \\ 0 & \quad\quad \text{otherwise.} \end{cases}$$

$$q_i^* = \begin{cases} q_i/||q_i||_i & \text{if} \quad q_i \neq 0 \\ e & \text{if} \quad q_i = 0 , \end{cases}$$

where e is a unit in O_i, and $||\cdot||_i$ is the induced norm in the current structure $(\, , \,)_i$. Let

$$x_{i+1} = x_i - \gamma r_i q_i^* ,$$

$$r_{i+1} = \beta r_i .$$

Introduce new coordinates O_{i+1}, with center x_{i+1}, radius r_{i+1} and containing hemispheres $U_i = \{x: ||x||_i \le r_i,$ $(x-x_i, q_i^*)_i \le 0\}$ in coordinates O_i. By construction $|W_{i+1}| = \beta^n r_i^n |W_1| = r_{i+1}^n |W_1|$. After $M = (\ell n\, \beta^{-1})^{-1} \ell n \frac{1}{\omega}$ iterations if $\min_i f_1(x_i) > \omega V_1$, the result is $*$; otherwise $x_{i_o} = \operatorname{argmin}\{f_o(x_i) | f_1(x_i) \le \omega V_1\}$.

<u>Theorem 5</u>: The ellipsoidal algorithm for convex problems has relative $*$-defect $\omega^*(B^M, C) \le \omega$.

<u>Proof</u>: Let $U_i^o = \overline{(W_i - U_i)}$.
Then $f_{j(i)}(x) \ge f_{j(i)}(x_i)$, $\forall x \in U_i^o$. By construction $W_i \subseteq U_i^o \cup W_{i+1}$, $\forall i$. Since $G = W_1$, we have $G = W_1 \subseteq W_{m+1} \cup (\bigcup_{i=1}^{M} U_i^o)$. Let $H_o = G \setminus \bigcup_{i=1}^{M} U_i^o$. Clearly $H_o \subseteq W_{M+1}$. Thus

$$|H_o| \le |W_{M+1}| \le \beta^{Mn} |W_1| = \beta^{Mn} |G| .$$

By the choice of M, we have

$$\left(\frac{|H_o|}{|G|} \right)^{\frac{1}{n}} \le \omega .$$

Since by construction, for $y \in G \setminus H^o$, $f_1(y) \leq \omega V_1$, we get $f_o(y) \geq a$, where

$$a = \begin{cases} +\infty & \text{if} \quad j(i) \geq 1 \ , \forall i \\ \min\{f_o(x_i) \mid j(i) = 0\}, & \text{otherwise.} \end{cases}$$

Thus the conditions of Lemma 1 hold. So we have the required result.

Now, we will transform relative $*$-defect to absolute $*$-defect. Let $V = \max\{V_o, V_1\}$. Clearly

$$V \leq R \cdot d ,$$

where R is radius of the initial sphere, and

$$d = \max\{\sup_{x \in G}\{|u| : u \in \partial f_1(x)\}, \sup_{x \in G}\{|u| : u \in \partial f_o(x)\}\}.$$

Then given $\varepsilon > 0$, choose ω such that

$$\omega V \leq \varepsilon.$$

Note that $\beta \leq e^{-1/2n^2}$ [see the next section).
Thus, in order to get the ε defect we need

$$M(\varepsilon) = 2n^2 \log(\frac{Rd}{\varepsilon}) = 2n^2 \lceil \log R + \log d + \log \frac{1}{\varepsilon} \rceil .$$

We say $\bar{x}$ is ε-optimal for P_o if

$$f(\bar{x}) - f(x^*) \leq \varepsilon, \ s(\bar{x}) - s(x^*) \leq \varepsilon$$

where x^* solves P_o.

Les us restate the above theorem in slightly different terminology.

Theorem 6: The ellipsoidal algorithm will give an ε-optimal solution to P_o in $M(\varepsilon)$ iterations.

3.7 SHOR'S 1977 PAPERS

After the publication of Judin-Nemirovskii's 1976 papers, Shor reconsidered his family of GGSD with certain parameters specified by Judin and Nemirovskii. He observed that the resulting algorithm was indeed the ellipsoidal algorithm. He gave the analytic formulae explicitly, and a slightly different interpretation of the algorithm. In his second paper in 1977 he calculated the rate of convergence of the algorithm for general unconstrained convex minimization. In this section we will review these two papers ([S18,S19]).

Ellipsoidal algorithm: analytic development

Let $\beta = (\frac{n-1}{n+1})^{\frac{1}{2}}$, $\alpha = \beta^{-1}$, $\sigma = (n^2/n^2 - 1)^{\frac{1}{2}}$. (1)

Consider the following convex programming problem:

$$\min_{x \in R^n} f(x).$$

Let x^* be a minimizer, and suppose that there is a positive constant R s.t. $x^* \in \{x: ||x - x_o|| \leq R\}$ for some arbitrary but known point x_o.

The algorithm

0. $x_o \in R^n$, $B_o = I_n$, $h_o = R/(n+1)$

1. At the k'th step we have $x_k \in R^n$, B_k, $h_k > 0$; for the k+1'th step we compute

$$g(x_k) = g_f(x_k) \in \partial f(x_k) . \tag{2}$$

   If $g(x_k) = 0$, stop; x_k is a minimizer.

2. Otherwise let

$$s_k = B_k^t g_f(x_k)/||B_k^t g_f(x_k)||.$$

3. $x_{k+1} = x_k - h_k \; B_k \; s_k$.
4. $B_{k+1} = B_k \; R_\beta(s_k)$.
5. $h_{k+1} = h_k \cdot \sigma$.
6. $k = k + 1$; go to 1.

Let $H_k = B_k B_k^t$, and $H_k' = h_k'^2 \; H_k$, $h_k' = h_k(n + 1)$. It is easy to verify that for $R = 2^L$, the above algorithm is identical to the algorithm presented by Gács-Lovász [G13], with A_k replaced by H_k' and the Shor algorithm presented by Skokov [S36].

Geometric description

Let us describe the algorithm geometrically. Consider $\phi_o = S_o = \{x: ||x - x_o|| \le R\}$. We evaluate $g_f(x_o)$. If it is zero, we are done; x_o is a global minimizer of f. Otherwise we form the hyperplane $< g_f(x_o), x - x_o > = 0$. We know that $x^* \in \bar{S} \equiv S_o \cap \{x: < g_f(x_o), x - x_o > \le 0\}$. We form the minimum volume ellipsoid around $\bar{S}$. The new ellipsoid is completely described by its volume and the condition that it contains $\bar{S}$. The center of the new ellipsoid is $x_1 = x_o - h_1 \; s_o$, where $s_o = g_f(x_o)/||g_f(x_o)||$. Now by dilating the space in the direction s_o: in the new space the image of ϕ_1, will be $\{y: ||y - y_1|| \le R\sigma\}$, where $y_1 = R_\alpha(s_o)x_1$, i.e. we have a sphere of radius $R\sigma$ around y_1. Now we apply the algorithm in the new space, and so on.

Lemma 1: [Shor 18] The sequence $\{x_k\}_{k=1}^{\infty}$ generated by the above algorithm satisfies the inequality:

$$||A_k(x_k - x^*)|| \le h_k' = h_k(n + 1) = R\sigma^k \;, \qquad (3)$$

where $A_k = B_k^{-1}$.

Proof: By induction on k. For k = 0, (3) becomes $||x_k-x^*|| \le R$, which is true by assumption.

Suppose (3) holds up to index k. Let $z_k = A_k(x_k-x^*)$. Then:

$$||z_{k+1}||^2 = ||A_{k+1}(x_{k+1}-x^*)||^2 = ||R_\alpha(s_k)A_k(x_k-x^*-h_kB_ks_k)||^2$$

$$= ||R_\alpha(s_k)(z_k-h_ks_k)||^2 = ||z_k-h_ks_k + (\alpha-1) < z_k-h_ks_k, s_k>s_k||^2$$

$$= ||z_k + (-\alpha h_k + (\alpha-1) <z_k,s_k>)s_k||^2$$

$$= ||z_k||^2 - 2h_k <z_k,s_k> \alpha^2 + (\alpha^2-1) <z_k,s_k>^2 + \alpha^2 h^2$$

$$= ||z_k||^2 - 2h_k <z_k,s_k> \frac{n+1}{n-1} + \frac{2}{n-1} <z_k,s_k>^2 + \frac{n+1}{n-1} h_k^2 .$$

Note that:

$$<z_k,s_k> = \frac{1}{||B_k^t g(x_k)||} <A_k(x_k-x^*), B_k^t g(x_k)>$$

$$= \frac{1}{||B_k^t g(x_k)||} <x_k-x^* , g(x_k)> \ge 0.$$

Hence

$$\frac{2}{n-1} <z_k,s_k>^2 - 2h_k \frac{n+1}{n-1} <z_k,s_k> = \frac{2}{n-1} [<z_k,s_k> \{<z_k,s_k> -h_k(n+1)\}] .$$

Since $||z_k|| \le h_k(n+1)$ by induction, and $||s_k|| = 1$,

$$<z_k,s_k> -h_k(n+1) \le ||z_k|| \, ||s_k|| - h_k(n+1)$$

$$= ||z_k|| - h_k(n+1) \le 0.$$

Hence

$$\frac{2}{n-1} \langle z_k, s_k \rangle^2 - 2h_k \frac{n+1}{n-1} \langle z_k, s_k \rangle \le 0. \tag{5}$$

Thus

$$||z_{k+1}||^2 \le ||z_k||^2 + \frac{n+1}{n-1} h_k^2 \le (n+1)^2 h_k^2 + \frac{(n+1)}{n-1} h_k^2$$

$$= (n+1)^2 \frac{n^2}{n^2-1} h_k^2 = (n+1)^2 h_{k+1}^2 . \qquad \square$$

Let $\phi_k = \{x: ||A_k(x_k - x)|| \le R\,\sigma^k = h_k'\}$. (6)

Clearly ϕ_k is an ellipsoid and equivalent to

$$\{x: (x-x_k)^t \frac{A_k^t}{h_k'^2} A_k (x-x_k) \le 1\} =$$

$$= \{x: (x-x_k)^t (H_k')^{-1} (x-x_k) \le 1\} . \tag{7}$$

To calculate $v(\phi_k)$, the volume of ϕ_k, consider the image of ϕ_k under A_k. Let $y = A_k x$, $y_k = A_k x_k$, and

$$\phi_k' = A_k \phi_k = \{A_k x: x \in \phi_k\} = \{A_k x: ||A_k x - A_k x_k|| \le h_k'\}.$$

(8)

$$= \{y: ||y - y_k|| \le h_k'\} .$$

Obviously, by calculus methods one has

$$v(\phi_k') = v(\phi_k) \det A_k = R^n \sigma^{nk} .$$

Thus

$$\frac{v(\phi_{k+1})}{v(\phi_k)} = \frac{v(\phi_{k+1}')}{\det A_{k+1}} \frac{\det A_k}{v(\phi_k')} = \frac{\det A_k}{\det A_k \det R_\alpha(s_k)} \frac{R^n (\sigma^{k+1})^n}{R^n (\sigma^k)^n}$$

$$= \beta\sigma^n = q_n ,$$

where $q_n = \beta\sigma^n = (\frac{n-1}{n+1})^{\frac{1}{2}} (\frac{n^2}{n^2-1})^{n/2} = \frac{n}{n+1} (\frac{n^2}{n^2-1})^{(n-1)/2}$. (9)

Lemma 2: $(\frac{n+1}{n-1})(\frac{n^2-1}{n^2})^n \geq e^{1/n}$ for $n \geq 2$

Proof: Let $f(x) = \frac{x+1}{x-1} (\frac{x^2-1}{x^2})^x$, $g(x) = e^{1/x}$, and let $f_1(x) = \log f(x)$, $g_1(x) = \log g(x)$, and $H(x) = f(x) - g(x)$. Since logarithmic function is monotonic, $f(x) \geq g(x)$ if and only if $f_1(x) \geq g_1(x)$. Direct verification shows that

$$\frac{d}{dx} (f_1(x) - g_1(x)) \leq 0, \quad \forall x > 1.$$

It is also routine to verify that, by L'hôpital's rule, $\lim_{x\to\infty} H(x) = 0$. $\frac{d}{dx} (f_1(x) - g_1(x)) \leq 0$ implies that $\frac{d}{dx} (f(x) - g(x)) \leq 0$, i.e. $\frac{dH(x)}{dx} \leq 0$. Together with $\lim_{x\to\infty} H(x) = 0$ we conclude that $H(x) \geq 0$. □

Corollary 3: $q_n = \beta\sigma^n = (\frac{n-1}{n+1})^{\frac{1}{2}} (\frac{n^2}{n^2-1})^{n/2} \leq e^{-1/2n}$. □

We would like to give a proof of some of the above claims.

For convenience let $R = 1$, $x_o = 0$; let us parametrize our calculations with respect to h. Without loss of generality we may assume $g_f(x_o) = - e_1$. Then

$$x_1 = +he_1 ,$$
$$y_1 = R_\alpha(e_1)x_1 = \alpha he_1 ,$$
$$y = R_\alpha(e_1)x .$$

Since $x = e_1 \in \bar{S}$, $||R_\alpha(e_1)e_1 - y_1|| \le \sigma$, $\Longrightarrow$

$$\alpha(-h+1) \le \sigma ,$$

and for $x \in S_o \cap \{x: x_1 = 0\}$, we have

$$R_\alpha(e_1)x = x .$$

Thus

$$||x - \alpha h e_1|| \le \sigma , \quad \Longrightarrow$$

$$\alpha^2 h^2 + ||x||^2 = 1 + \alpha^2 h^2 \le \sigma^2 .$$

Now, the volume of $\phi_1 = \alpha^{-1} \sigma^n = \beta\sigma^n$. We would like to minimize

$$\alpha^{-1}\sigma^n = F(h)$$

s.t.

$$\alpha(1-h) \le \sigma$$

$$1 + \alpha^2 h^2 \le \sigma^2 .$$

Without loss of generality we may assume that at the minimum the inequalities become equalities. Then we can solve for α, σ as

$$\alpha = \sqrt{1 - 2h} ,$$

$$\alpha^2 = \frac{(1-h)^2}{1-2h} .$$

Then $F(h) = \sigma^n \alpha^{-1}$ becomes

$$F(h) = (1-h)^n \left(\frac{1}{1-2h}\right)^{(n-1)/2}$$

$$F'(h) = 0 \Rightarrow h = \frac{1}{n+1} .$$

Thus

$$\sigma = \sqrt{\frac{n^2}{n^2-1}} ,$$

$$\alpha = \sqrt{\frac{n+1}{n-1}} .$$

Convergence rate of the algorithm

The above algorithm in each iteration decreases the volume of ϕ_k, and thus the volume of the set of candidate points minimizing f. The decrease is geometric with ratio $q_n = \beta\sigma^n < e^{-1/2n}$.

It is equally true that $\rho_k = \min_{1\le\ell\le k} f(x_\ell) - f(x^*)$ has a geometrical convergence rate which is almost $q_n^{1/n}$ (in fact $\exp\{-\frac{1}{4n^2}\}$).

Lemma 4: $\frac{1}{x}\left(\frac{n+1}{n-1}\right)^{x/n}\left(\frac{n^2-1}{n^2}\right)^x \ge e^{x/2n^2}$ for $x \ge x_o$ for some x_o.

Proof: Let $f(x) = \frac{1}{x}\left(\frac{n+1}{n-1}\right)^{x/n}\left(\frac{n^2-1}{n^2}\right)^x$, $g(x) = e^{x/2n^2}$,

$f_1(x) = \log f(x)$, $g_1(x) = \log g(x) = \frac{x}{2n^2}$, and let

$$H_1(x) = f_1(x) - g_1(x) = -\log x + x \log \bar{q} - \frac{x}{2n^2} ,$$

where

$$\bar{q} = \left(\frac{n+1}{n-1}\right)^{1/n}\left(\frac{n^2-1}{n^2}\right).$$

Thus

$$H_1'(x) = \frac{d}{dx} H_1(x) = -\frac{1}{x} + \log \bar{q} - \frac{1}{2n^2} .$$

By Lemma 2 $\bar{q} \ge e^{1/n^2}$, hence $\log \bar{q} \ge \frac{1}{n^2}$. So for $x \ge 2n^2$, $H_1'(x) \ge 0$. But

$$H_1(2n^2) = -\log 2n^2 + 2n^2 \log \bar{q} - 1 \ge 1 - \log 2n^2.$$

For $x \geq 2n^2 = x_o$, we have

$$H_1(x) = \int_{x_o}^{x} H_1'(t)dt \geq H_1(x_o) \geq \int_{2n^2}^{x} (-\frac{1}{t} + \frac{1}{2n^2})dt + 1 - \log 2n^2.$$

Hence

$$H_1(t) \geq (\frac{t}{2n^2} - \log t)\Big|_{x_o}^{t} + 1 - \log 2n^2$$

$$= \frac{t}{2n^2} - \log t = H_2(t).$$

In order to complete the proof, it suffices to show that $H_2(x) \geq 0$ for some $x \geq 2n^2$.

Let $U(t) = t - \log 2t$, then $U(1) = 1 - \log 2 \geq 0$, and $U'(t) = 1 - \frac{1}{t} \geq 0$, for $t \geq 1$ $\therefore$ $U(t) \geq 0$, for $t \geq 1$.

In fact, let $a = \log 2n^2$, and $t_o = 2a\, 2n^2$.

$$H_2(t_o) = 2a - a - \log 2a = a - \log 2a \geq 0$$

since

$$a \geq 1.$$

Hence

$$H_1(t) \geq H_2(t) \geq 0 \text{ for } t \geq \tilde{t} \text{ for some } \tilde{t} \leq t_o.$$

Thus

$$f_1(x) \geq g_1(x) \Rightarrow f(x) \geq g(x). \qquad \square$$

<u>Theorem 5</u>: [Shor] For the ellipsoidal algorithm, assume

$$||g_f(x_\ell)|| \leq d \quad \forall \ell .$$

Then

$$\min_{1 \leq \ell \leq k} ||g_{\Psi_\ell}(y_\ell)|| \leq d \frac{\sqrt{k(\alpha^2-1)}}{\sqrt{1-\beta^{2k/n}}} \beta^{k/n}, \quad \forall k \geq k_o, \qquad (10)$$

where $\Psi_k(y_k) = f(x_k)$.

Proof: Let us write A_k as a product of an orthogonal matrix O_k and a symmetric positive definite matrix S_k; i.e.

$$A_k = O_k \cdot S_k ,$$

$$S_k = \sqrt{A_k^t A_k} , \quad O_k = A_k S_k^{-1} .$$

Since

$$A_k = R_\alpha(s_{k-1}) \dots R_\alpha(s_o), \text{ and } O_k \text{ orthogonal}$$

$$\det A_k = \det S_k = \prod_{i=1}^{k} \det R_\alpha(s_{i-1}) = \alpha^k . \tag{11}$$

Let

$$\lambda_1^k \le \lambda_2^k \dots \le \lambda_n^k ,$$

where $\{\lambda_i^k\}$ $i = 1,\dots,n$ are the eigenvalues of S_k, in non-decreasing order, and $\{e_i^k: i = 1,\dots,n\}$ is the corresponding system of normalized eigenvectors

$$S_k e_i^k = \lambda_i^k e_i^k .$$

Then

$$\prod_{i=1}^{n} \lambda_i^k = \alpha^k , \quad \lambda_n^k \ge \alpha^{k/n} . \tag{12}$$

Let $\bar{e}_i^k = O_k e_i^k$. Since O_k is orthogonal, $\{\bar{e}_i^k : i = 1,\dots,n\}$ is also orthonormalized. Let us expand $g_f(x_k)$ as

$$g_f(x_k) = \sum_{i=1}^{n} g_i^k e_i^k . \tag{13}$$

Since

$$||g_f(x_k)|| \le d, \text{ we have}$$

$$\sum_{i=1}^{n} (g_i^k)^2 \le d^2 . \tag{14}$$

Since

$$A_k = O_k \cdot S_k , \quad B_k^t = (A_k^{-1})^t = O_k \cdot S_k^{-1} .$$

Thus

$$\tilde{g}_k = B_k^t \, g_f(x_k) = O_k \cdot S_k^{-1} \Sigma g_i^k \, e_i^k = O_k \, \Sigma \frac{g_i^k}{\lambda_i^k} \, e_i^k$$

$$= \sum_i \frac{g_i^k}{\lambda_i^k} \, \bar{e}_i^k \quad . \tag{15}$$

Furthermore, by definition

$$S_{k+1} = \frac{\tilde{g}_k}{||g_k||} = \sum_{i=1}^{n} \mu_i^k \, e^{-k} , \tag{16}$$

where

$$\mu_i^k = \frac{g_i^k}{\lambda_i^k ||\tilde{g}_k||} \quad .$$

Suppose, contrary to the assertion of theorem, we have $\forall r, \; r \le k$

$$U_k \le \min_{r \le k} ||\tilde{g}_r|| > d \, \frac{\sqrt{k(\alpha^2-1)}}{\sqrt{\alpha^{2k/n}-1}} \quad . \tag{17}$$

Then we have

$$|\mu_i^r| < \frac{|g_i^r| \sqrt{\alpha^{2k/n}-1}}{\lambda_i^r \, d \sqrt{k(\alpha^2-1)}} , \quad \forall r \le k.$$

We will estimate $(\lambda_n^{r+1})^2 - (\lambda_n^r)^2$, where λ_n^r is the maximum eigenvalue of the operator S_r. Consider an arbitrary vector $a \in R^n$, $||a|| = 1$; we have

$$a = \sum_{i=1}^{r} a_i \, e_i^r , \quad \Sigma a_i^2 = 1, \quad A_r a = \sum_{i=1}^{r} a_i \lambda_i^r \, \bar{e}_i^k \quad .$$

Then

$$\lambda_n^{r+1} = \max_{||a||=1} ||A_{r+1}\, a|| = \max_{||a||=1} ||R_\alpha(s_{r+1})A_r\, a||.$$

But we have by Lemma 5.2

$$||R_\alpha(s)x||^2 = ||x||^2 + (\alpha^2-1) < x,s >^2 .$$

Letting $x = A_r a$, we have

$$(\lambda_n^{r+1})^2 = \max_{||a||=1} \left[\sum_{i=1}^{n} a_i^2(\lambda_i^r)^2 + (\alpha^2-1)\left(\sum_{i=1}^{r} a_i\lambda_i^r\mu_i^r\right)^2\right]$$

$$\leq \max_{||a||=1} \left[(\lambda_n^r)^2 + (\alpha^2-1)\left(\sum_{i=1}^{n} |a_i g_i^r|\right)^2 \cdot \frac{\alpha^{2k/n}-1}{k\cdot d^2(\alpha^2-1)}\right]$$

$$\leq (\lambda_n^r)^2 + \frac{\alpha^{2k/n}-1}{k} , \text{ by using (14).}$$

Thus, we have

$$(\lambda_n^{r+1})^2 < (\lambda_n^r)^2 + \frac{\alpha^{2k/n}-1}{k} \quad \text{for} \quad r \leq k. \tag{18}$$

Writing down the above inequality for $r = 0,\ldots,k$, and cancelling similar terms, and noting $\lambda_n^0 = 1$, we have

$$(\lambda_n^k)^2 < 1 + k \cdot \frac{\alpha^{2k/n}-1}{k} = \alpha^{2k/n}$$

i.e. $(\lambda_n^k) < \alpha^{k/n}$, a contradiction to (12).

Thus the assertion holds. □

Note that we did not use anything about the properties of function f. The theorem was originally stated for an unconstrained function. However, it is obvious from the proof that it holds whenever g(x) is a quasi-gradient or any valid cut separating the solution points from the rest.

Theorem 6: Under the above assumption, and $||x_o - x^*|| \le R$, we have

$$\rho_k = \min_{0 \le \ell \le k} [f(x_\ell) - f(x^*)] \le Rd \sqrt{\frac{k(\alpha^2-1)}{1-\alpha^{2k/n}}}\, q_n^{k/n} .$$

Proof: Let the minimum for Theorem 5 occur at ℓ_o. Then

$$\rho_k = \min_{0 \le \ell \le k} f(x_\ell) - f(x^*) = f(x_{\ell_o}) - f(x^*) = \Psi_o(y_{\ell_o}) - \Psi(y^*)$$

$$\le \langle g_\Psi(y_{\ell_o}), \; y_{\ell_o} - y^* \rangle \le ||g\,(y_{\ell_o})||\; ||y_{\ell_o} - y^*|| \le$$

$$\textcircled{1} \le d\sqrt{\frac{k(\alpha^2-1)}{1-\beta^{2k/n}}}\; \beta^{k/n}\, R \cdot \sigma^{\ell_o}$$

$$\textcircled{2} \le \sqrt{\frac{\alpha^2-1}{1-\beta^{2/n}}}\; R \cdot d\,\sigma^k\, \beta^{k/n} \quad \sqrt{k}$$

$$\le \sqrt{\frac{\alpha^2-1}{1-\beta^{2k/n}}}\; \sqrt{k}\, R \cdot d\, q_n^{k/n} .$$

(1) follows from Theorem 5, and $y^* \in \phi_{\ell_o} = \{y: ||y - y_{\ell_o}|| \le R \cdot \sigma^{\ell_o}\}$.

(2) follows from the fact $\sigma > 1$, hence for $\ell_o \le k$, $\sigma^{\ell_o} \le \sigma^k$.

Corollary 7: $\rho_k \le C_o \cdot R \cdot d \exp\{-\frac{k}{4n^2}\}$.

Proof: Follows from Lemma 4, where $C_o = \sqrt{\frac{\alpha^2-1}{1-\beta^{2k/n}}}$ (for k large enough $C_o \le 1$).

As we stated earlier in the proof of Theorem 5, all we need is the fact that g(x) is a valid cut separating the solution points from the rest. Thus we have

Theorem 8: In the ellipsoidal algorithm under the above hypothesis, we have

$$\min_{1\le \ell \le k} ||B_\ell^t\, g(x_\ell)|| \le \frac{d\sqrt{k(\alpha^2-1)}}{\sqrt{1-\beta^{2k/n}}}\,\beta^{k/n}, \quad \text{for } k \ge k_o. \qquad \square$$

Imitating part of the proof of Theorem 6, and combining with Lemma 4, we have

Theorem 9: Let $||x_o-x^*|| \le R$, $||g(x)|| \le d$, $\forall x \in E_o$, and $g(x) \in G(x)$ is a valid cut. Then, for $M(\varepsilon) = 4n^2 \log \frac{Rd}{\varepsilon}$, the ellipsoidal algorithm will generate an $\bar{x}$ such that

$$|\langle g(\bar{x}), \bar{x} - x^* \rangle| \le \varepsilon \qquad \square$$

Let $k = M(\varepsilon)$, and ℓ_o be defined by Theorem 8. When one chooses $g(x) \in G(x)$, where

$$G(x) = \begin{cases} \partial f(x) & s(x) \le \varepsilon \\ \partial s(x) & s(x) > \varepsilon, \end{cases}$$

then the above theorem generalizes Theorem 6 to constrained optimization. Sometime after obtaining this generalization, we have realized that this generalization is in fact contained in the main theorem of Judin-Nemirovskii, i.e. Theorem 3.6. However, we will use it in Chapter 4 for a direct algorithm. We will also use it in Chapter 6, for which the Judin-Nemirovskii result does not apply.

PART B. HACHIYAN'S CONTRIBUTION AND THE PROOF OF THE MAIN RESULT

3.8 HACHIYAN'S CONTRIBUTION

In this section we aim to clarify Hachiyan's contribution in proving LP $\in \mathcal{P}$.

Evidently Hachiyan realized the potential of the ellipsoidal algorithm in deciding whether LP $\in \mathcal{P}$. He obtained a simple characterization in terms of an estimate of the length of data to write the problem, and he showed that one can implement the algorithm on a Turing Machine.

One can identify three problems related to a linear inequality system:

(i) Decision Problem (D): Decide whether $P = \{x: Ax \leq b\}$ is non-empty or not. This is a yes or no question. If the answer is yes, that is $P \neq \phi$, we do not require an $x \in P$. This is the approach which the computer scientist takes.

(ii) Feasibility Problem (LI): Given $P = \{x: Ax \leq b\}$ it is required to find an $x \in P$, or statement that $P = \phi$, possibly a proof of the latter.

(iii) Optimization Problem (LP): This is the usual linear programming problem; $\max \{cx \mid x \in P\}$.

It is well-known that the above problems are polynomially equivalent in the sense that if one has an algorithm for one of them, by calling that algorithm as a subroutine or oracle in polynomial number of times one can solve the other problem. Equivalence of (LI) and (LP) is the content of LP duality. Since LI is stronger than D, it suffices to show that one can solve LI using D. First suppose $m \geq n$, and $r(A) = n$. Then starting with the first inequality, we set $I = \phi$ and define

$$P_i = \{x: a_j x = b_j, \quad j \in I,$$
$$a_i x = b_i$$
$$a_k x \leq b_k, \quad k > i\}.$$

If $P = \phi$, stop. Otherwise

Do for $i = 1, \ldots, m$

if $P_i \neq \phi$, then $I = I \cup \{e\}$

end.

If $P \neq \phi$, $m \geq n$, it has at least one basic solution.
The solution $a_i x = b_i$, $i \in I$ is a solution to LI.
If $m < n$, then any basic solution of the resulting system is a solution to IL.

One can solve LP, using D, directly, i.e. without using LP duality. For this, one first checks whether $P = \phi$. If $P = \phi$, we are done. Then one should check whether LP has a finite optimum. Since by the fundamental theorem of linear programming, if LP has a (finite) optimum, then it has a basic point attaining the optimum, we have an a priori upper bound on the value of the optimum objective value. One can ask the question, is $P' = \{x: Ax \geq b,\ cx > \gamma\} = \phi$, γ being the apriori optimum ($\gamma = 2^L||c||$ will do). If $P' \neq \phi$, then LP is unbounded. Otherwise we know that a finite optimum exists. By using a binary search, we reduce the size of the interval in which the optimum value lies. Using the rationality because of integrality of A, b, c, we know that the optimum value is a rational number of the form $z_o + p/q$, p and q are relatively prime, and $p < q$, $z_o \in Z$. We can compute an upper bound for q a priori, using the encoding of P. Let $\varepsilon = 1/2q^2$. Then in $\log 2\gamma/\varepsilon$ binary search we localize the interval I_ε in which the optimum value lies, with $|I_\varepsilon| \leq \varepsilon$. Furthermore, from the theory of continued fractions (Hinchin [H16], Reiss [R3]), we know that in the interval I_ε there is only one rational number of the form s/q, $s \leq q$. Thus expanding the best known value of the objective function which is known to be in I_ε, we obtain the exact value of optimum of LP, say $\bar{z}$. Then any (basic) solution of $P_2 = \{x: Ax \leq b,\ cx \geq \bar{z}\}$ will be a solution to LP.

Length of the input

It is now well accepted that (see Garey and Johnson [G15], Aho et al [A 4], Karp [K 7, K 8], Cook [C29], Tarjan [T2]), from the point of view of computational complexity, one considers the problems as a set of strings of {0,1} for which the algorithm-computer-Turing Machine-RAM answers as yes or no.

For the decision problem associated with $P = \{x: Ax \leq b\}$, the question is whether P is non-empty or not. We measure the size of the input for P as the total space needed to store the coefficients of A and b in binary form. Binary encoding is well accepted; for one thing it resembles most of the digital computers which work in binary digits, for another it is equivalent, i.e. same order, with any k-ary encoding for $k \geq 2$.

For an integer a, we need $\pi_1(a)$ places to write a in binary encoding, without taking care of sign, where

$$\pi_1(a) = \begin{cases} 1 & a = 0 \\ 1 + \lfloor \log_2|a| \rfloor & a \neq 0 . \end{cases} \tag{1}$$

We need, in general, one more space for the sign. Thus

$$\pi(a) = \begin{cases} 2 & a = 0 \\ 2 + \lfloor \log_2|a| \rfloor & a \neq 0 . \end{cases} \tag{2}$$

So the length of such an encoding of $P = \{x: Ax \leq b\}$ is

$$\begin{aligned} L_1 &= \pi(m) + \pi(n) + \sum_{i,j} \pi(a_{ij}) + \sum_j \pi(b_j) \\ &= (2 + \lfloor \log m \rfloor) + (2 + \lfloor \log n \rfloor + (2mn + \sum_{a_{ij}\neq 0} \lfloor \log|a_{ij}| \rfloor \\ &\quad + (2m + \sum_{b_j \neq 0} \lfloor \log|b_j| \rfloor). \end{aligned} \tag{3}$$

Note for $x \geq 1$

$$\log(x+1) = \log x + \int_x^{x+1} \frac{1}{t}\,dt \le \log x + \log 2.$$

Thus, for $a_{ij} \ne 0$, $a_{ij} \in Z$,

$$\log(|a_{ij}| + 1) \le \log|a_{ij}| + \log 2.$$

Since $\log_2 x = \dfrac{\log x}{\log 2}$,

$$\log_2(x+1) = \frac{\log(x+1)}{\log 2} \le \frac{1}{\log 2}[\log x + \log 2]$$

$$= \log_2 x + 1.$$

Thus $\pi'(x) = \log_2(x) \le \pi(x)$, for $x > 0$.

Let us recall that Hachiyan used the following as the length of the input:

$$L = \lfloor \sum_{i,j} \log_2(|a_{ij}| + 1) + \sum_j \log_2(|b_j| + 1) + \log_2 mn \rfloor + 1. \tag{4}$$

Clearly, $L \le L_1$. In this sense L is a lower bound on the size of the length of the input.

Use of L

Although the above definition of the length of the input is well-accepted in general†, no one before Hachiyan has tried to use such a measure of encoding to obtain a characterisation of some feature of P. What Hachiyan did was to use L in an a priori explicit characterization in locating and in determining a measure of inconsistency for the decision problem.

A simple use of such an L is:

Lemma 1: Let A' be a submatrix of [A,I,b]. If det A' ≠ 0, then $1 \le |\det A'| \le 2^L/mn$.

---

† There are some objections. See [C13], [M20], [T7].

Proof: The lower bound follows from the fact that A, b are integer valued. The upper bound follows from Hadamard's theorem. But we will give a brute force derivation, which also has some combinatorial flavour. Without loss of generality we assume that A' is an n × n matrix. By definition we have

$$\det A' = \sum_{\sigma\in S_n} \operatorname{sgn} \sigma \prod_{i=1}^{n} a'_{i\sigma(i)}$$

$$\le \sum_{\sigma\in S_n} \prod_{i=1}^{n} |a'_{i\sigma(i)}|.$$

Now, for $t \in R^k$, we define

$$f_t(x) = \prod_{i=1}^{k} (1 + t_i x) = \sum_{\ell=0}^{k} x^\ell \Big(\sum_{|I|=\ell} \prod_{s\in I} t_s \Big) = \sum_{\ell=0}^{k} c_\ell x^\ell ,$$

$$I \subset \{1,\ldots,k\}.$$

For $k = n^2$, $t \equiv (|a'_{ij}|)$, as a vector in R^{n^2}, we have

$$\sum_{\sigma\in S_n} |a'_{i\sigma(i)}| \le c_n \le f_t(1) \le \frac{2^L}{mn} .$$ Thus

$$|\det A'| \le \frac{2^L}{mn} .$$

We would like to point out that the restriction of integrality of A, b resulted in the lower and upper bounds of the above inequality in Lemma 1. If we remove the integrality restriction, we may still get an upper bound by Hadamard's theorem, but we no longer have a lower bound. It turns out that the lower bound is very important.

As a simple consequence of the above lemma, one can restrict himself to an euclidian ball of radius 2^L (in fact $\frac{2^L}{m\sqrt{n}}$ if $m \ge n$).

Lemma 2: If $P \neq \phi$, then for a vertex x of P one has

$$|x_i| \leq \frac{2^L}{mn} .$$

Proof: By Cramer's theorem, x_i can be written as

$$x_i = \frac{|D^i|}{|D|}$$

where D, D^i are appropriate submatrices of [A,I,b], and det $D \neq 0$. Since $|\det D| \geq 1$, we have

$$|x_i| \leq |D^i| \leq \frac{2^L}{mn} . \qquad \square$$

If $P \neq \phi$, and does not have any vertex, then $P \cap Q_i \neq \phi$, and has vertices, for some orthant Q_i of R^n. Since $P \cap Q_i \subseteq P$, we have the following

Corollary 3: If $P \neq \phi$, then $P \cap \{x: ||x|| \leq 2^L\} \neq \phi$. $\square$

Hachiyan used L in obtaining a characterization of non-solvability of P. If P is empty, i.e. the system of linear inequalities is inconsistent, one may measure the degree of inconsistency as follows: let

$$\theta(x) = \max_i a_i x - b_i ,$$

and

$$\theta(x^*) = \inf_{x \in R^n} a_i x - b_i .$$

$$\theta(x^*) \leq 0 \iff P \neq \phi.$$

If $\theta(x^*) > 0$, $P = \phi$, i.e. system is inconsistent. Let

$$b_i' = b_i + \theta(x^*), \quad \text{then}$$

$$P' = \{x: Ax \leq b'\}$$

is consistent.

This measure of inconsistency was previously introduced by Chernikov [C20], Eremin [E12]. Chernikov [C21] gave an explicit formula for $\theta(x^*)$ in terms of the value of the determinants of the augmented system [A,I,b]. They used the term 'defect'.

<u>Lemma 4</u>: If $P = \phi$, then $\theta(x) \geq 2\cdot 2^{-L}, \quad \forall x \in R^n$.

<u>Proof</u>: We will give several slightly different proofs.

(i) By Chernikov's result [C20,C21,E12],

$$\theta(x^*) = \frac{\Delta}{\Delta_1+\cdots+\Delta_n},$$

and since $\Delta \geq 1$, $\Delta_i \leq \frac{2^L}{mn}$,

$$\theta(x^*) \geq m\cdot 2^{-L} \geq 2\cdot 2^{-L}$$

(ii) Consider the following LP:

$$\min \quad t$$

$$a_i x - b_i \leq t.$$

Since $\theta(x^*) = t > 0$, (because $P = \phi$), we have by the fundamental theorem of linear programming, a basis A_1 satisfying

$$t = \frac{\det[A_1,e,b_1]}{\det[A_1,e]},$$

where b_1 is the corresponding part of b, e is a vector of 1's. Since

$$|\det[A_1,e]| \leq \sum_i |\det[A_1,e_i]| \leq n \cdot \frac{2^L}{mn} = \frac{2^L}{m},$$

and

$$|\det[A_1,e,\ b_1]| \geq 1,$$

$$t \geq m \cdot 2^{-L} \geq 2 \cdot 2^{-L}$$

(iii) Since x^* is a minimizer of $\theta(x)$, then $\underline{0} \in \partial\theta(x^*)$. But that means

$$(*) \quad \begin{cases} \lambda_i \geq 0 \quad , \quad \sum_{i\in I} \lambda_i = 1 \\ \sum_{i\in I} a_i\lambda_i = 0 \ , \end{cases}$$

$$a_i x^* - b_i = \theta(x^*) = t, \quad i \in I.$$

Since $a_i x^* = b_i + t$, for $i \in I$,

$$0 = \sum_{i\in I} \lambda_i a_i x^* = \sum_{i\in I} \lambda_i (b_i + t) = \sum_{i\in I} \lambda_i b_i + t.$$

Since $t > 0$, we have

$$\sum_{i\in I} \lambda_i b_i = -t.$$

Since any basic solution for (*) has the form

$$\lambda_i = \frac{z_i}{z} \quad , \text{ for } \lambda_i > 0, \quad z_i > 0, \quad z \leq \frac{2^L}{m} \quad \text{and}$$

$$t = - \Sigma \lambda_i b_i \geq m\, 2^{-L} \geq 2 \cdot 2^{-L}.$$

Approximate arithmetic

Formally one defines class P as problems polynomially solvable on a deterministic Turing Machine. Hachiyan also assumed that the Turing Machine has a capacity to record a 'number', i.e. he assumed that if a is written on a tape of the machine then i) $|a|$ is bounded by a polynomial of the length of the input, ii) it has an upper bound on the number of binary points after the decimal, which also depends on the length of the input.

Subsequently he has shown that the above conditions are satisfied. i.e. i) the numbers do not grow too fast, that is they are bounded by an a priori bound depending on L, ii) we can do approximate calculations satisfying a certain accuracy which will enable us to verify the validity of the algorithm.

We represent rational numbers during the course of the algorithm in their binary expansion, not in quotient form. For real numbers, we approximate them by a rational number with an accuracy which we will specify later. That is, we will choose a $\delta > 0$, $\delta = \delta(L)$, a priori, and perform the calculations to an accuracy of δ.

Let us set some notation. We will represent an ellipsoid as $E = (x,B) = \{x + By: ||y|| \leq 1\}$. For $x \in R^n$, $||x||$ will be the usual euclidean norm. For a $n \times n$ matrix B,

$$||B|| = \sup_{||x||=1} ||Bx||,$$

is the operator norm. It is well-known that $||B||$ is equal to the square root of the largest eigenvalue of B^tB.
Let us note that Hachiyan uses the Frobenius norm, i.e. $||B||_F^2 = \sum_{i,j} b_{ij}^2$. By λE we mean the ellipsoid $(x,\lambda B)$ $=\{x + \lambda By : ||y|| \leq 1\}$. For $E = (x,B)$, and $E' = (x',B')$ we say E' approximates E with accuracy δ if

$$||E-E'|| \equiv ||x-x'|| + ||B-B'||_F \leq \delta$$

which means the sum of errors in approximations of center and matrix B is less than or equal to δ, which implies that each one is less than δ, and which in turn guarantees that in each component of the matrix $|B-B'|$ is less than or equal to δ.

For such an approximation consider the image of a particular point z, with $||z|| \le 1$. Let

$$y = x + Bz \ , \quad y' = x' + B'z \ .$$

Then,

$$||y-y'|| = ||x-x' + (B-B')z|| \le ||x-x'|| + ||B-B'|| \ ||z|| \le \delta .$$

Hence, every point of E' is at most δ away from a point of E, and vice versa. In particular the same is true for boundary points. Thus we have

$$d(y,E') \le \delta, \quad d(y',E) \le \delta, \quad d(E,E') \le \delta \ ,$$

where

$$d(A,B) = \inf \{d(x,y) \,|\, x \in A, \ y \in B\}$$

is the Hausdorff metric.

Let us define the thickness of A, r(A), for a compact set as the size of the largest euclidean ball contained in A. For ellipsoids r(E) represents the size of the smallest semi-axis, or, equivalently, $\sqrt{\ }$ of the smallest eigenvalue of BB^t.

Lemma 5: Let E = (x,B) be a non-degenerate ellipsoid with thickness r(E), and suppose E' = (x',B') approximates E = $(x, \lambda B)$ with accuracy δ, where $\lambda \in (1,2)$ and

$$\delta \le (\lambda-1) r(E) \ .$$

Then E' contains E; hence E' is non-degenerate. Moreover mes E'/mes E $\le (2\lambda-1)^n$ where mes A is the volume of A in R^n .

Proof: Consider the centers x and x'. Now

$$||x-x'|| \le \delta \le (\lambda-1)r(E)$$

$$\Rightarrow \quad x' \in E, \quad x \in E' .$$

Note that $d(E,E_\lambda) = (\lambda-1)r(E) = \rho$, and by assumption $\delta \le \rho$. By definition $d(E_\lambda,E') \le \delta$. It suffices to show that $\partial E' \cap \text{int } E = \phi$, for verifying $E \subset E'$. Since

$$d(\partial E, \partial E') \ge d(\partial E, \partial E_\lambda) - d(\partial E_\lambda, \partial E') = \rho - \delta \ge 0,$$

we have $\partial E' \cap \text{int } E = \phi$, proving $E \subset E'$. For the second assertion we need to show that

$$E' \subseteq (2\lambda-1)E.$$

Let $\bar{\lambda} = 2\lambda-1$, $\bar{\lambda} \in (1,3)$. Since $x' \in E$, we have $x' \in E_{\bar{\lambda}}$. Since $d(E_\lambda, E_{\bar{\lambda}}) = (\lambda-1)r(E) = \rho$, and $\delta \le \rho$,

$$d(E',E_{\bar{\lambda}}) \ge d(E_{\bar{\lambda}},E_\lambda)-d(E_\lambda,E') = \rho - \lambda \ge 0,$$

showing that $E' \subseteq E_{\bar{\lambda}}$. Then

$$\frac{\text{mes } E'}{\text{mes } E} \le \frac{\text{mes } E_{\bar{\lambda}}}{\text{mes } E} = \frac{\det[(2\lambda-1)B]}{\det B} = (2\lambda-1)^n.$$

Let us recall the basic iteration of the ellipsoidal algorithm. For $q \in \partial f(x_k)$, $g \ne 0$:

$$x^+ = x - \frac{Bs}{(n+1)},$$

$$B^+ = \sigma B^t R_\beta(s),$$

where

$$s = B^t g/||B^t g||.$$

Then

$$||x^+|| \le ||x|| + \frac{||B||}{n+1} ,$$

and

$$||B^+|| \le \sigma||B|| \; ||R_\beta(s)|| = \sigma||B|| .$$

Thus

$$r(E^+) = \min_{||z||=1} ||B^+z|| = \min \left\{ \min_{\substack{||z||=1 \\ (z,s)=1}} ||B^+z||, \; \min_{\substack{||z||=1 \\ (z,s)=0}} ||B^+z|| \right\}$$

(since norm is uniformly convex). Now

$$\min_{\substack{||z||=1 \\ (z,s)=1}} ||B^+z|| = \min_{||z||=1} \sigma\beta||Bz|| \ge \sigma\beta r(E) = \frac{n}{n+1} r(E) ,$$

$$\min_{\substack{||z||=1 \\ (z,s)=0}} ||B^+z|| = \min_{\substack{||z||=1 \\ (z,s)=0}} \sigma||Bz|| \ge \sigma r(E) > \frac{n}{n+1} r(E) .$$

For arbitrary $||z|| = 1$, let $z = z_1 + z_2$, $\langle z_1, z_2\rangle = 0$, $\langle s, z_2\rangle = 0$, then $B^+z = \sigma\beta R_\beta(z) = \sigma\beta(\beta z_1 + z_2)$. Since

$$||Bz|| \ge r(E)||z|| , \forall z ,$$

$$||B^+z|| = \sigma||B(\beta z_1 + z_2)|| \ge \sigma r(E)||\beta z_1 + z_2||.$$

Since for the above choice of z_1, z_2, $\beta\epsilon(0,1)$,

$$||\beta z_1 + z_2|| \ge \beta,$$

$$||B^+z|| \ge \sigma\beta r(E) = \frac{n}{n+1} r(E) .$$ □

We need the following inequality.

Lemma 6: For $n \geq 3$, the following holds:

$$\sigma\lambda_n + \frac{1}{24n^2} \leq 2^{1/n^2} ,$$

where $\lambda_n = 1 + \dfrac{1}{16n^2}$.

Proof: Consider the left hand side; since

$$\sigma = \sqrt{\frac{n^2}{n^2-1}} = \sqrt{1 + \frac{1}{n^2-1}} \leq (1 + \frac{1}{2(n^2-1)}), \text{ then}$$

$$LH \leq (1 + \frac{1}{16n^2})(1 + \frac{1}{2(n^2-1)}) + \frac{1}{24n^2}$$

$$= 1 + \frac{1}{16n^2} + \frac{1}{24n^2} + \frac{1}{2(n^2-1)} + \frac{1}{32n^2(n^2-1)} .$$

Since

$$\frac{1}{n^2-1} \leq \frac{9}{8}\frac{1}{n^2} ,$$

$$LH \leq 1 + \frac{1}{n^2}(\frac{1}{16} + \frac{1}{24} + \frac{9}{16}) + \frac{3}{64}\frac{1}{n^4}$$

$$\leq 1 + (\log_e 2)\frac{1}{n^2} + \frac{(\log_e 2)^2}{2!}\frac{1}{n^4} \leq \exp(\frac{\log_e 2}{n^2})$$

$$= \exp(\log_e 2^{1/n^2}) = 2^{1/n^2} . \qquad \square$$

Lemma 7: $(2\lambda_n-1)^n q_n \leq 2^{1/2n}$.

Proof: We know that $q_n \leq \exp\{-\frac{1}{2n}\}$ (By Corollary 3.7.3). Let $\mu_n = 2\lambda_n - 1 = 1 + \dfrac{1}{8n^2}$. Then it suffices to show that $\mu_n^n \exp\{-\frac{1}{2n}\} \leq 2^{1/2n}$.

Multiplying both sides with $\exp\{\frac{1}{2n}\}$, we have

$$\mu_n^n = (1 + \frac{1}{8n^2})^n \leq (2e)^{1/2n} . \qquad (+)$$

Since

$$\log(1 + \frac{1}{8n^2})^n = n \log(1 + \frac{1}{8n^2})$$

and

$$\log(2e)^{1/2n} = \frac{1}{2n}(1 + \log_e 2).$$

Thus (+) is equivalent to

$$2n^2 \log(1 + \frac{1}{8n^2}) \le 1 + \log_e 2 .$$

Since

$$\log(1 + x) < x , \quad \text{for} \quad x \in (0,1)$$

we have

$$2n^2 \log(1 + \frac{1}{8n^2}) < \frac{2n^2}{8n^2} = \frac{1}{4} < 1 + \log 2 . \qquad \square$$

<u>Lemma 8</u>: Let $E = (x,B)$ be a non-degenerate ellipsoid with thickness $r(E)$, $E^+ = (x^+,B^+)$ be the next ellipsoid obtained by exact arithmetic. Furthermore let $E' = (x',B')$ approximate $E_\lambda = (x^+, \lambda_n B^+)$ with accuracy δ, where

$$\delta \le \frac{r(E)}{24n^2} .$$

Then $E^+ \subseteq E'$,

$$||x'|| \le ||x|| + \frac{||B||}{n} , \quad ||B'|| \le ||B|| 2^{1/n^2} ,$$

and $\quad \text{mes } E'/\text{mes } E \le 2^{-1/2n}$.

<u>Proof</u>: For the inclusion, by Lemma 5, it suffices to show that

$$\delta \le (\lambda_n - 1) r(E^+).$$

But

$$\delta \le \frac{r(E)}{24n^2} \le \frac{n+1}{n}\,\frac{r(E^+)}{24n^2} \le \frac{3}{2}\,\frac{1}{24n^2}\,r(E^+) = (\lambda_n - 1)r(E^+).$$

We know that $||x^+ - x'|| \le \delta$, $||x^+ - x|| \le \frac{1}{n+1}\,||B||$,

$$B^+ = \sigma B R_\beta(s) \Rightarrow ||B^+|| \le \sigma||B||\,,\quad ||\lambda_n B^+ - B'|| \le \delta\,.$$

Thus,

$$||x'|| \le ||x|| + \frac{1}{n+1}||B|| + \frac{r(E)}{24n^2} \le ||x|| + ||B||\left(\frac{1}{n+1} + \frac{1}{24n^2}\right)$$

$$\le ||x|| + \frac{||B||}{n}\,,$$

$$||B'|| \le \lambda||B^+|| + \delta \le \lambda\sigma||B|| + \frac{r(E)}{24n^2} \le ||B||\left(\lambda_n\sigma + \frac{1}{24n^2}\right)$$

$$\le 2^{1/n^2}\,||B||.$$

Since $\delta \le (\lambda_n - 1)r(E^+)$,

$$\frac{\text{mes } E'}{\text{mes } E^+} = (\lambda_n - 1)^n\,,$$

$$\frac{\text{mes } E'}{\text{mes } E} = \frac{\text{mes } E'}{\text{mes } E^+}\cdot\frac{\text{mes } E^+}{\text{mes } E} = (\lambda_n - 1)^n\, q_n \le 2^{-1/2n}$$

by Lemma 8. □

The description of the algorithm is complete, except the total number iterations, and an a priori bound for δ. Since we do not know $r(E)$ in each step, we need to use an a priori δ as mentioned above. We will choose $\delta = 2^{-2L}/24n^2$. 2^{-2L} is the thicknes of the ellipsoid on which we conclude the result of the decision problem. In other words it comes from the stopping criteria which we will study in the next section.

Let us conclude this section by giving, once more, the basic step of the algorithm in approximate arithmetic.

$$x' \approx x - \frac{Bs}{n+1} ,$$

$$B' \approx (1 + \frac{1}{16n^2}) B \, R_\beta(s),$$

$$s \approx B^t g/||B^t g||,$$

where $\approx$ denotes approximate arithmetic. All the calculations are performed with accuracy $\delta = 2^{-2L/24n^2}$.

By assumption the tape of the Turing Machine has 3L (binary) places after the decimal point which allows us to do arithmetic with δ accuracy.

Let us note that Hachiyan in his Doklady paper [H 2], gave a different perturbation; he used $2^{1/8n^2}$ instead of $(1+ \frac{1}{16n^2})$. The above development closely follows the recent paper [H 4]. Gács-Lovász [G13], Stone [S41], Padberg-Rao [P4], Grötschel-Lovász-Schrijver [G48] also analyzed or gave provisions for the approximate arithmetic. We will study Ursic's [U1] result in Chapter 4.

3.9 PROOF OF THE MAIN RESULT

In this section we will review various proofs of Hachiyan's main result: LP $\in$ P. We will give his proof, together with some well-known, but different proofs which are incorrectly identified with Hachiyan's proof. We also will deduce the main result from earlier works of Judin-Nemirovskii and Shor as explained previously.

We will assume exact arithmetic. Since it is shown in the previous sections that the approximate arithmetic will only have the effect of increasing q_n to $\tilde{q}_n$ (incidentally both are less than $2^{-1/2n}$), we do not lose any generality.

Let us recall some of the earlier definitions. By <u>decision problem</u> (D) we mean to decide whether $P = \{x \in R^n : Ax \le b\} = \phi$ or not. By <u>feasibility problem</u> (LI) we mean (D) with the additional property that when $P \ne \phi$, we require an $x \in P$. LP speaks for itself; i.e. $\max\{cx | x \in P\}$.

Since all these three problems are polynomially equivalent, it suffices to show that any one of them belongs to $\mathcal{P}$.

Hachiyan in his Doklady paper [H 2], (see also [H 4]) showed that $D \in \mathcal{P}$. But for this, it suffices to decide whether $\theta(x^*) = \min_{x \in \mathcal{R}^n} \theta(x)$, where $\theta(x) = \max_i \{a_i x - b_i\}$, is less than $2 \cdot 2^{-L}$ or not.

As a corollary to Judin-Nemirovskii result

Let us represent the decision problem as $\vec{f} \in C(G, \mathcal{R}^n, 1, V_0, V_1)$, where

$$f_0(x) = 0, \qquad (1)$$

$$f_1(x) = \theta(x).$$

We know that, by definition of L, $R < 2^L$, $d < 2^L$. In order to solve D it suffices to find a point x such that $f_1(x) \le 2^{-L}$. Thus by choosing $R = d = 2^L$, $\varepsilon = 2^{-L}$, we obtain in $M = 6n^2L$ (in fact $(\log_e 2)6n^2L$) iterations an ε-optimal solution for the above problem. If during the algorithm we generate an x such that $f_1(x) \le 2^{-L}$, we answer the decision problem positively, i.e. $P \ne \phi$, otherwise $P = \phi$.

Naturally one can interchange the roles of f_0 and f_1. Then let $\bar{\theta} = \min\{\theta(x_\ell) : \ell \le M\}$. If $\bar{\theta} \le 2^{-L}$, then P is nonempty. Otherwise $\bar{\theta} \ge 2 \cdot 2^{-L}$, and we know that $\theta(x^*) > 0$ (since we guarantee the inequality $\bar{\theta} - \theta(x^*) \le \varepsilon$), and hence proving P is empty.

As a corollary of Shor's work

Let us consider the unconstrained minimization problem

$$\min_{x \in R^n} \theta(x). \tag{2}$$

By virtue of the localization lemma, we may consider the problem

$$\min_{x \in E_o} \theta(x), \tag{3}$$

where E_o is the initial ellipsoid, i.e. euclidean ball of radius 2^L centered at zero.

Then, by Corollary 3.7, we have

$$\min_{\ell \in k} [\theta(x_\ell) - \theta(x^*)] \le r.d. \exp\{-\frac{k}{4n^2}\} .$$

Letting $R = d = 2^L$, then in $k = K = 12n^2L$ iterations of the ellipsoid algorithm applied to problem (3), we obtain $\bar{x}$ such that

$$\theta(\bar{x}) - \theta(x^*) \le 2^{-L},$$

where $\bar{x} = \text{argmin}\{\theta(x_\ell) | \ell \le K\}$. Then, if $\theta(\bar{x}) \le 2^{-L}$, we have $\theta(x^*) \le 0$, implying P is non-empty. Otherwise $\theta(\bar{x}) \ge \theta(x^*) \ge 2 \cdot 2^{-L}$, and P is empty.

We should note that recently Goffin [G35], obtained a slightly sharpened version of the convergence result of Shor, and Judin-Nemirovskii for unconstrained convex minimization, from which a similar proof follows immediately.

Hachiyan's proof

Hachiyan considers the minimization problem (4). He chooses his cuts from the subgradient set of $\theta(x)$. Let us denote a_{i_k} as the subgradient used at iteration k.

Let $P_k = \{x: a_{i_k}(x - x_k) > 0\}$. It is the useless part. Obviously $\theta(x) > \theta(x_k)$, $\forall x \in P_k$. Let $S = E_o$. Let us define a few more sets:

$$S_k = S \cap [\bigcup_{\ell=0}^{k} P_\ell] , \qquad D_k = S \setminus S_k . \tag{4}$$

At iteration k exactly all the candidate points are contained in D_k, and $D_k \subseteq E_{k+1}$.

Definition: The thickness r(A) of a set A is the radius of the largest euclidean sphere that can be placed in A, i.e.

$$r(A) = \max_{x \in A} \min_{y \in \partial_A} ||x - y|| , \tag{5}$$

where ∂_A is the boundary of A. □

For $N = 6n^2L$,

$$\det B_N = \text{vol } (E_N) \le 2^{nL}(2^{-1/2n})^N = 2^{-2nL} . \tag{6}$$

Lemma 1: $r(D_t) \le r(E_{t+1}) \le 2^{-2L}$, for some $t \le N$. □

Clearly $x^* \in D_t$. Let

$$V = \{x: ||x - x^*|| \le r_2\}, \quad r_2 = 4 \cdot 2^{-2L} \tag{7}$$

$$W = V \cap D_t . \tag{8}$$

Then the boundary of W, ∂_W can be written as

$$\partial_W = \partial_S \cup \partial_V \cup \partial_{\tilde{P}} , \tag{9}$$

∂_S : formed by sphere $\{x: ||x|| = 2^L\}$

∂_V : formed by sphere $\{x: ||x - x^*|| = r_2\}$

$\partial_{\tilde{P}}$: formed by union of boundaries of P_k's.

Lemma 2: Let U_1, U_2 be euclidean balls of radius $r_1 > r_2 > 0$. If the center of U_2 is in U_1, then

$$r(U_1 \cap U_2) \geq r_2/2.$$

Proof: Let x be center of U_2. Let y be such that $d(x,y) = d(x, \partial U_1)$. If $x \neq y$, let $z = x + r/2(x - y)$. If $x = y$, then let e be the inward normal to U_1 at x, and let $z = x + r/2\ e$. Then $U = \{u: ||z - u|| \leq r/2\} \subset U_1 \cap U_2$. □

Lemma 3: $\partial_{\tilde{P}} \neq \phi$.

Proof: Otherwise $W \subset V \cap S$, i.e. $W = V \cap D_t = V \cap S$. But $r(W) \geq r_2/2 = 2\cdot 2^{-2L}$, by Lemma 2, contradicting the fact that

$$r(W) \leq r(D_t) \leq r(E_{t+1}) \leq 2^{-2L} . \tag{10}$$

Thus V intersects at least one hyperplane, say $a_{ik}(x - x_k) = 0$. Let $y \in S$ be an arbitrary point in the intersection, i.e.

$$a_{ik}(x_k - y) = 0, \ ||x^* - y|| \leq r_2 = 4\cdot 2^{-2L} .$$

Since $y \in \partial S_t$, $y \in \partial P_k$ for some k. Hence

$$\theta(y) \geq \theta(x_k) \geq \theta_{t+1} = \min_{\ell \leq t+1} \theta(x_\ell) . \tag{11}$$

Indeed $\theta(x)$ is Lipschitz with constant $\max_i ||a_i|| < \frac{2^L}{mn}$. It follows that

$$|\theta(x^*) - \theta(y)| \leq \frac{2^L}{mn} ||x^* - y|| \leq \frac{2^L}{mn} 4\cdot 2^{-2L} \leq 2^{-L} .$$

Then

$$\min_{x \in D_t} \theta(x) = \theta(x^*) \geq \theta(y) - 2^{-L} \geq \theta_{t+1} - 2^{-L} .$$

Thus we have established that

$$\theta(\bar{x}) - \theta(x^*) \leq 2^{-L} \text{ , where}$$

$$\theta(\bar{x}) = \min_{\ell \leq N} (x_\ell).$$

This settles the decision problem.

In all of these proofs, the basic technique is to bound the variation of $\theta(x)$ on the last ellipsoid which is considered. In fact, in Shor's development, it is not necessarily the last one, but this is a minor point.

It is worth repeating that Hachiyan settled the decision problem by the above arguments, not the feasibility problem. Apparently he did not need a full dimensionality hypothesis.

The remaining proofs of the result that we will review, are all, roughly, based on the ingenious, but ad-hoc, method of Gács-Lovász.

<u>Other proofs</u>

Hachiyan's Doklady paper presented all the essential ideas for solving the decision problem. He explicitly states that through the last ellipsoid one bounds the variation of $\theta(x)$, that is difference between $\theta(x^*)$ and the best current value $\theta(\bar{x})$ is $\leq 2^{-L}$. Partly because he does not say how to obtain this bounding and the upper bound on the number of iterations, and partly the fact that Shor's 77 papers, and Judin-Nemirovskii's 76 papers, either were not available and/or not so easily understandable, the ellipsoidal algorithm and the main result was not understood for some time until the Gács-Lovász paper.

Gács-Lovász solved a slightly different problem. They showed that the ellipsoidal algorithm, when applied to

$\bar{P} = \{x \in R^n: Ax < b\}$, solves the feasibility problem in polynomial time. Key to their proof is the following:

<u>Lemma 4</u>: If $\bar{P} \neq \phi$, then

$$\text{vol}(\bar{P} \cap \{x: ||x|| < 2^L\}) \leq 2^{-(n+1)L}. \qquad \square$$

When $\bar{P} \neq \phi$, since it is open, it has a positive volume. They calculated the volume of $\bar{P}$ in the initial ellipsoid by constructing a simplex in solution set, and by computing its volume.

In order to solve the feasibility problem for P, they suggested to solve $P' = \{x: Ax < b + 2^{-L} \cdot \bar{1}\}$. The equivalence of two problem is guaranteed by the following

<u>Lemma 5</u>: P has a solution $\Longleftrightarrow$ P' has. $\square$

Gács-Lovász provided a polynomial algorithm in order to obtain a solution x of P from x' of P'. Their algorithm can be described, equivalently, first by obtaining a basic point $\tilde{x}$ of P', and by using the hyperplanes defining $\tilde{x}$ in P', thus obtaining an x (which is also a basic point) in P.

There is one more difference between Hachiyan, and Gács-Lovász versions. Hachiyan, by minimizing $\theta(x)$, has to take as cuts the subgradient of $\theta(x)$. Whereas Gács-Lovász used a stronger structure, it allowed them to take any valid cut using the defining inequalities.

Since Gács-Lovász almost identified their proof with that of Hachiyan's, the ellipsoidal algorithm, especially the Gács-Lovász version, has been known as 'Hachiyan's Algorithm' for some time. It is interesting to note that even the perturbation is referred to Hachiyan incorrectly.

An $x \in P' \setminus P$ is called an 'approximate' solution of P, and referred to Hachiyan by some authors, e.g. Pickel [P 9], Charnes et.al [C 13].

Bland-Goldfarb-Todd [B 24], Padberg-Rao [P 3], Aspvall-Stone [A 20] gave refined versions of the Gács-Lovász approach. The same goes for Murty [M 30], and Rinnoy-Kan et.al [R 4], etc.

Korte-Schrader [K 19] gave a slightly different approach. Their proof depends on metrics and norms defined by a symmetric positive matrix representing the ellipsoid in each iteration. It is equivalent to the observation that thickness (or inner radius) of successive ellipsoids are bounded above by a geometric ratio (n'th root of the normalized volume). Since they also solve the strict inequality system, at the last ellipsoid the current thickness will be smaller than the thickness of the simplex in the solution set. Hence a proof of emptiness of the solution set.

4 Linear programming via ellipsoidal algorithm

The ellipsoidal algorithm has become famous for its application in resolving the question whether LP $\in P$. In this chapter we will discuss various ways of using the ellipsoidal algorithm in solving linear programming problems. We suggest a direct ellipsoidal algorithm for linear programming. We discuss the idea of combining the simplex method with the ellipsoidal algorithm.

We present an accelerated direct ellipsoidal algorithm and give its performance on some well-known hard LP problems. The basic idea can be adapted to handle the linear complementarity problem (LCP). We discuss the solution of some well-known hard LCP problems with the accelerated algorithm.

The ellipsoidal algorithm gives in polynomial time an ε-optimal solution to any convex programming problem, hence to a linear program. The linear structure of the underlying polyhedron and integrality restrictions allow us to find the exact solution of the LP from an ε-optimal solution with a polynomial effort, for a suitable ε. This partly manifests itself in the coefficients of the inverse of any basis. We introduce a class of matrices with certain restrictions on inverse matrices. This class contains totally unimodular matrices. We compute the complexity of a direct ellipsoidal algorithm for this class of linear programs.

4.1 AS HACHIYAN DESCRIBED

Hachiyan used the ellipsoidal algorithm for the decision problem, and the decision problem for the linear programming problem. He used the decision problem polynomial number of times to solve an LP. Let us set the linear programming problem explicitly:

$$\text{LP} \qquad \begin{array}{l} \max \ cx \\ Ax \le b \ . \end{array}$$

Let us denote $P = \{x: Ax \le b\}$. One first checks whether P is non-empty or not. If P is empty, we are done; LP is infeasible. Otherwise one should check whether LP is unbounded or not. By linear programming duality theory, LP in bounded if and only if Dual LP: $\{\min yb: A^T y = c,\ y \ge 0\}$ is feasible, i.e. $\{y \in R^n_+ : A^T y = c\} \ne \phi$. This can be settled by the ellipsoidal algorithm. Let us define the length of the input of LP as

$$L_1 = L + \lfloor \sum_i \log_2(|c_j| + 1) \rfloor \ .$$

If LP is bounded, then $\psi^+ = cx^*$ is a rational number of the form $\psi^+ = t/s$, t and s are relatively prime integers, and $|t| \le 2^{L_1}$, $|s| \le 2^{L_1}$. One can approximate the value of ψ^+ to any desired accuracy by binary search. One starts with the interval $[-2^{L_1}, 2^{L_1}]$. We can represent $\psi^+ = [\psi^+] + t'/s'$, $0 \le t' < s'$, t', and s' relatively prime. In $\lceil \log \frac{2^{L_1+2}}{\varepsilon} \rceil$ binary search, i.e. solving the decision problems of the form: is P_k empty, where

$$P_k: \qquad \begin{array}{l} Ax \quad b \\ cx \ge \psi_k \end{array}$$

where ψ_k is the midpoint of $[\alpha_k, \beta_k]$. If P_k is empty let $\alpha_{k+1} = \alpha_k$, $\beta_k = \psi_k$, otherwise $\alpha_{k+1} = \psi_k$, $\beta_{k+1} = \beta_k$. One can obtain an approximation $\tilde{\psi}$ to ψ^+ satisfying $|\tilde{\psi} - \psi^+| \le \varepsilon$.

When $\varepsilon \le \frac{1}{2 \cdot 2^{2L_1}}$, in particular for $\varepsilon = 2^{-2L_1-2}$, by the theory of continued fractions one can compute ψ^+ exactly. The convergent of ψ^+ which is in the form $\tilde{t}/\tilde{s}$, with maximal $\tilde{s}$ satisfying $\tilde{s} \le 2^{L_1}$ will be ψ^+ (see Hinchin [H16], Reiss [R 3]).

Since $\tilde{\psi}$ has $2L_1 + 2$ binary digits after the decimal point, i.e. $O(L_1)$, in $O(L_1)$ arithmetic one can compute Ψ^* over $O(L_1)$ binary digits.

After determining ψ^+, any solution to

$$P': \quad \begin{aligned} Ax &\le b \\ cx &\ge \psi^+ \end{aligned}$$

will be an optimal solution to LP. Hachiyan suggests to find a solution to P' by determining an compatible equality system, for which any solution to the equality system is a solution to P'. This done in m' , the number of inequalities, steps. Each inequality, in turn, is set to an equality to see whether its intersection with the solution set of P' is empty or not. If it is empty, then it is superfluous, thus it can be thrown away. Otherwise we set the inequality as equality permanently. Thus in m' steps we throw away some of the inequalities and keep the rest of the inequalities as equalities. Let us call the remaining system U. After eliminating the dependent rows, one should adjoin, if necessary, trivial inequalities. This can be done by adjoining the trivial equalities $x_j = 0$, in succession to the current system, and checking whether the adjoined system is consistent. If the rank of U is r, then by adjoining the appropriate (n-r), $x_j = 0$, we end up with a new system of equalities, say U', with a unique solution. A solution to U' can be obtained in $O(n^3)$ iteration by well-known techniques.

By linear programming duality theory, any solution (x,y) to

$$P'': \quad \begin{aligned} Ax &\le b \\ A^T y &= c \\ cx &\ge by \\ y &\ge 0 \end{aligned}$$

will give an optimal solution for primal and dual LP. Hachiyan suggested the solution of P" via solution of associated decision problems, as described in the previous paragraph.

Hachiyan also suggested to solve the general PL via simple LP's. Since his algorithms are based on the ellipsoidal algorithm, he basically solves decision problem. His idea is solve the following linear programs:

$$LP_j': \quad \begin{aligned} &\min x_j \\ &Ax \le b, \end{aligned}$$

where $Ax \le b$ includes $cx = cx^*$, $x_k = x_k^*$, $k \le j$ which are obtained in the previous LP's. The reason behind the algorithm is the possibility of having better performance of the new algorithm.

We should once more point out that the reason that Hachiyan was suggesting the above versions of the algorithm, which are rather expensive, is that he did not have a perturbation lemma.

One of the biggest achievements of the Gács-Lovász paper [G13] was the perturbation lemma. In other words, they observed that given any solution to $Ax < b + \bar{2}^L \cdot \bar{1}$, one can obtain a solution to $Ax \le b$, with a polynomial work. That is, any point which proves that P is non-empty, i.e. $f(x) \le \bar{2}^L$, also gives way to a point in P.

Gács-Lovász proof of the perturbation lemma was through a polynomial algorithm. Essentially it generates another point x' such that $I(x) \subset I(x')$, where $I(x) = \{i: a_i x \ge b_i\}$. When I(x) is maximal, hyperplanes in I(x) define a vertex which is a solution.

Lemma 1: [G13] $a_i x \le b_i$, $\forall i$ has a solution if and only if $a_i x < b_i + 2^{-L}$, $\forall i$ has. □

J.F. Maurras introduced the idea that when one is near enough to a solution to a linear system, then just by moving to a basic solution via a simplex type argument, one can obtain the (optimal) solution of the problem at hand.

We now give a proof of Lemma 1, which is based on ideas of [M13]. Lemma 1 theoretically guarantees finding a solution to P, when in fact one has solved a perturbed problem.

Let $\bar{x}$ be a solution to $a_i x < b_i + 2^{-L}$, $\forall i$. Let us introduce slack variables so that we have

$$A\bar{x} + \bar{y} = b + \epsilon \cdot \bar{1},$$
$$y \ge 0$$

where $\varepsilon = 2^{-L}$, $\bar{1} = (1,1,\ldots,1)$. Let $(\tilde{x},\tilde{y})$ be a basic solution of the perturbed system, with a basis B, obtained from $(\bar{x},\bar{y})$. Such a basic solution can be obtained by the algorithm VERTEX with polynomial effort, i.e. n simplex pivots (see Chapter 2). Then

$$(\tilde{x},\tilde{y})_B = B^{-1}(b+\varepsilon \cdot \bar{1}).$$

Let us define

$$(\underset{\sim}{x},\underset{\sim}{y})_B = B^{-1}b.$$

It suffices to show that $(\underset{\sim}{y})_B \ge 0$, for $\underset{\sim}{x}$ a solution to $Ax \le b$. Note that

$$\tilde{y}_i - y_i = \epsilon B^{-1}_{i\cdot} \cdot \bar{1} .$$

Noting that

$$B^{-1}_{ij} = \frac{D_{ij}}{D}, \quad |D_{ij}| < \frac{2^L}{mn}, \quad D < \frac{2^L}{2mn}$$

$$|\tilde{y}_i - y_i| = |\varepsilon \sum_j \frac{D_{ij}}{D}| < \varepsilon \cdot \frac{2^L}{n \cdot D} = \frac{1}{n \cdot D} .$$

On the other hand if

$$y_i < 0 \Rightarrow |y_i| \geq \frac{1}{D} .$$

Since $\tilde{y}_i \geq 0$ we have a contradiction.

Once one has the perturbation lemma, it is quite natural to use binary search with the feasibility problem (LI) to solve LP. This is done by Padberg-Rao [P 3], independent of Hachiyan. They used some elaborate techniques, including deep cut (we will discuss some of the acceleration techniques later in the chapter).

4.2 A DIRECT ELLIPSOIDAL ALGORITHM

In their original development of the ellipsoidal algorithm for general constrained convex minimization, Judin and Merirovskii introduced, essentially, ε-feasibility and ε-optimality. That is, when the current point satisfies $s(x) \leq \varepsilon$, they used an objective function cut. They also stated that without the ε-perturbation of the constraints, there is no finite algorithm within the class which contains Levin's algorithm and the ellipsoidal algorithm which solves the minimization problem for arbitrary precision (which satisfies the constraints).

When the current point is feasible, to use an objective function cut is quite natural, and suggested by Shor [S 19], Goldfarb-Todd [G 40], Pickel [P 9], and many others. Grötschel-Lovász-Schrijver [G48], in their fundamental paper on the consequences of the ellipsoidal algorithm introduced an ε-perturbation of the constraints. However, their concepts are not exactly the same as ours. They allow a certain arbitrariness in choosing an objective function cut or constraint cut, when the current point is ε-feasible (A discussion on an example in their paper will be given later in this section). Let us write the LP as

$$LP_1 \quad \begin{array}{l} \max f(x) = cx \\ \text{s.t. } s(x) \le 0 \,, \end{array}$$

where $s(x) = \max_{1 \le i \le m} \{a_i^t x - b_i\}$.

Let

$$G_\varepsilon(x) = \begin{cases} -\partial f(x) & , \quad s(x) \le \varepsilon \\ \partial s(x) & , \quad s(x) > \varepsilon \,. \end{cases}$$

We say $\bar{x}$ is <u>ε-optimal</u> for LP_1 if

$$f(x^*) - f(\bar{x}) \le \varepsilon,$$

and

$$s(\bar{x}) < \varepsilon,$$

where x^* solves LP.

Let x_{ℓ_o} be the point which satisfies:

$$\min_{\ell \le k} ||B_\ell^t \, g(x_\ell)|| \le \frac{d\sqrt{\alpha^2 - 1}}{\sqrt{1-\beta^{k/n}}} \, \beta^{k/n}$$

in Theorem 3.7.8, where $k = K(\varepsilon) = 4n^2 \log \frac{Rd}{\varepsilon}$.

For a basis B, let $f(B) = c_B B^{-1} b$ be the value of f at the basic solution corresponding to B.

<u>Lemma 1</u>: For (feasible) bases B_1 and B_2

$$f(B_1) \ne f(B_2) \Rightarrow |f(B_1) - f(B_2)| \ge mn2^{-L}$$

□

Now, for $\epsilon = 2^{-3L}$, let $K = K(\varepsilon)$ be such that

$$\min_{\ell \le K} ||B_\ell^t\, g(x_\ell)||\sigma^K \cdot R \le \varepsilon. \qquad (*)$$

Theorem 2: The ellipsoidal algorithm solves LP with at most $K = K(\varepsilon)$ iterations.

Proof: Let us note that

$$K = K(\varepsilon) = 4n^2(|\log 2^{-3L}| + \log 2^L + \log 2^L) = 20n^2L .$$

If the algorithm does not stop before $K = K(\varepsilon)$, then $\exists\ \ell_o,\ \ell_o \le K$, such that

$$|\langle B^T_{\ell_o} g(x_{\ell_o}),\ y_{\ell_o} - y^* \rangle| < \varepsilon .$$

Now, consider the value of $s(x_{\ell_o})$:

i) If $s(x_{\ell_o}) \le 0$, then $\exists$ basic feasible point $\underset{\sim}{x}$ in $\{x\colon Ax \le b\}$ obtained by algorithm VERTEX such that $f(\underset{\sim}{x}) \ge f(x_{\ell_o})$. Then

$$f(x^*) - \varepsilon < f(x_{\ell_o}) \le f(\underset{\sim}{x}) \le f(x^*) \Longrightarrow f(x\ ) - f(\underset{\sim}{x}) < \varepsilon$$

By Lemma 1, $f(\underset{\sim}{x}) = f(x^*)$ and we are done.

ii) If $0 < s(x_{\ell_o}) \le \varepsilon$, let $\tilde{x}$ be basic feasible point in $\{x\colon Ax \le b + \varepsilon\}$, with $f(\tilde{x}) \ge f(x_{\ell_o})$, obtained from algorithm VERTEX. Let $\underset{\sim}{x}$ be the corresponding basic point of $P = \{x\colon Ax \le b\}$. In other words $\exists$ a bases B such that

$$\tilde{x} = B^{-1}(b + \varepsilon \cdot 1)$$
$$\underset{\sim}{x} = B^{-1}b .$$

Since

$$|c(\tilde{x}-\underset{\sim}{x})| \le ||c_B\ B^{-1}\ \varepsilon \cdot \bar{1}\ || \le \varepsilon||c||\ \frac{2^L}{mn}\sqrt{n}\cdot\sqrt{n} < 2^{-L},$$

$$f(\underset{\sim}{x}) \ge f(\tilde{x})-2^{-L}\ .$$

We also have

$$f(\underset{\sim}{x}) \le f(x^*)\ ,\ f(x_{\ell_o}) \ge f(x^*)-2^{-3L} \ge f(x^*)-2^{-L}\ .$$

Then

$$f(\underset{\sim}{x}) \ge f(\tilde{x})-2^{-L} \ge f(x_{\ell_o})-2^{-L} \ge f(x^*)-2\cdot 2^{-L}$$

implying

$$|f(\underset{\sim}{x})-f(x^*)| \le 2\cdot 2^{-L}\ ,$$

by Lemma 1, we have $f(\underset{\sim}{x}) = f(x^*)$.

iii) If $s(x_{\ell_o}) > \varepsilon$, $g(x_{\ell_o}) \in \partial s(x_{\ell_o})$.

Hence $s(x_{\ell_o})-s(x^*) < \varepsilon$.

If LP is feasible, then $s(x^*) = 0$, thus $s(x_{\ell_o}) < \varepsilon$, contrary to assumption. So, if $s(x_{\ell_o}) > \varepsilon$, LP is infeasible.

□

Remarks

Although the algorithm takes $g(x) \in \partial s(x)$, when $s(x) > \varepsilon$, i.e. from the most violated ones, it suffices to take $g(x) = a_i$ such that $a_i x - b_i > \varepsilon$.

It should be pointed out that introduction of ε is due to the nature of the algorithm, and is independent of the full dimensionality of P. This is transparent in convex programming. In this regard we would like to comment on an example of Grötschel-Lovász-Schrijver. In [G48], they consider a class of full-dimensional convex bodies in R^2 and assume an oracle for a strong separation algorithm. They conclude that without the explicit knowledge of the ball contained in the convex body, there is no algorithm at all to solve the weak optimization problem. As a corollary, they also conclude a similar negative result for non-full dimensional bounded convex bodies. We would like to point out that the negative result comes from the restrictions on the oracle and the use of a strong separation algorithm. Let us remark that in their main theorem, Grötschel-Lovász-Schrijver use a weak separation algorithm. Their time bound depends on r, the radius of the ball contained in the convex body, explicitly, whereas that of ours does not. If one considers the problem $\{\max cx | x \leq +1, -x \leq -1\}$, in polynomial time one finds an ε-optimal solution, then by some additional work, e.g. via VERTEX (which is also polynomial), one obtains the required solution for a suitable ε . In general, as long as we have a (weak) separation algorithm, we can find an ε-optimal solution for a convex programming problem, in polynomial time in $|\log \frac{1}{\varepsilon}|$, for any $\varepsilon > 0$. When we have a linear and rational structure, we can obtain the exact solution (or strong optimizer) from an ε-optimal solution for a suitable ε. If the underlying polyhedron is irrational, we can not compute, at least currently, an a priori ε, such that when $P = \phi$, $s(x) \geq \varepsilon$, $\forall x \in R^n$, and a consequence of this, an ε that will allow us to obtain the exact solution of LP from an ε-optimal solution via VERTEX or similar polynomial-time algorithms.

4.3 THE COMPLEXITY OF THE DIRECT ALGORITHM FOR A CLASS OF LINEAR PROGRAMS

In the ellipsoidal algorithm L, the length of input, is used mainly to bound the size of the determinants of the linear inequality system. Of course one can replace L with log Δn, where Δ is the maximum determinant of the system [A,I,b]. This is observed, also, by Hachiyan [H 4] and Padberg and Rao [P 3]. In fact we are using the above information in bounding elements of inverse matrices.

We study the minimax characterization in terms of basic solution representations. This will give us estimates of the complexity of ellipsoidal algorithms much less than the straightforward application of Hachiyan' result.

Let R be such that

$$||x|| \le R, \quad \forall x \in P,$$

and

$$d = \max \{||a_i||_{i\in I} , ||c||\} ,$$

$$M(\varepsilon) = 2n^2 \log \frac{Rd}{\varepsilon} .$$

R determines the initial ellipsoid, and $g(x) \in G_\varepsilon(x)$ represents the cut used at point x. The following is a restatement of a well-known result, i.e. Theorem 3.6.6.

Theorem 1: The ellipsoidal algorithm gives an ε-optimal solution for LP in $M(\varepsilon)$ iterations with $g(x) \in G_\varepsilon(x)$.

□

Let C_q consist of all matrices A such that every basic solution to $Ax \le b$, for $A \in Z^{m\times n}$, $b \in Z^m$ has the form $x_i = \frac{n_i}{q}$, for $n_i \in Z$, $q \in Z^+$. Furthermore, for any basis B of A, it is assumed that $|(B^{-1})_{ij}| \le 1, \forall,j$.

Clearly C_q is non-void. For q = 1, it contains all t.u. systems. For q = 2 it contains some multicommodity flow problems, etc.

Lemma 2: If P is not solvable (i.e. empty), then

$$s(x) \geq \frac{1}{r'q}, \quad \forall x \in R^n ,$$

where r = r' - 1, and r is the rank of A.

Proof: Observe that $s(x) \leq 0$ if x is a solution to P. If P is not solvable then $s(x) > 0$, $\forall x \in R^n$. Consider the linear program

$$LP_1: \quad \begin{array}{l} \min\ t \\ a_i x - b_i \leq t \ , \ i = 1,\ldots,m. \end{array}$$

Since $s(x) > 0$, $\forall x \in R^n$, $t > 0$. Consider any basic point of LP_1 (optimal). There exist submatrices A_1, I_1 such that

$$A_1 x - e \cdot t_1 + I_1 y = b_1 ,$$

where $e = (1,1,\ldots,1)$, and $r(A_1) = r(A) = r$.
Using Cramer's rule,

$$t = \frac{\det[A_1,b_1,I_1]}{\det[A_1,e,I_1]} = \frac{|\det[A_1,b_1,I_1]|}{|\det[A_1,e,I_1]|} .$$

But $|\det[A_1,e,I_1]| \leq \sum_i |\det[A_1,e_i,I_1]| \leq r'|\det[A_1,I_1]|$

since $r(A_1,e,I_1) = r' \leq r + 1$. ($e_i$ is the usual i^{th} unit vector.) Thus,

$$t \geq \frac{1}{r} \frac{|\det[A_1,b_1,I_1]|}{|\det[A_1,I_1]|} \geq \frac{1}{rq} ,$$

since $\dfrac{|\det[A_1,b_1,I_1]|}{|\det[A_1,I_1]|} = \dfrac{k}{q}$, for some k.

<u>Corollary 3</u>: If $s(x) < \frac{1}{r'q}$ for some $x \in R^n$, then P is solvable.

Given $\bar{x}$ in $P = \{x: Ax \leq b\}$, one can, by using either the projection method or the simplex method, move to a vertex of P in at most n steps. One can also improve the objective function simultaneously via VERTEX.

<u>Lemma 4</u>: Let $\varepsilon = \frac{1}{2nq}$, let $\bar{x}$ be such that $s(\bar{x}) < \varepsilon$. Then a solution to P can be obtained by algorithm VERTEX.

<u>Proof</u>: Let $\bar{y} \geq 0$ such that

$$A\bar{x} + I\bar{y} = b + e \cdot \varepsilon . \tag{1}$$

Algorithm VERTEX will give a bases B, and a pair $(\tilde{x},\tilde{y})$, solutions to (1)

$$(\tilde{x},\tilde{y})_B = B^{-1}(b + \varepsilon \cdot e) .$$

Let us define $\underset{\sim}{x}$, $\underset{\sim}{y}$ as:

$$(\underset{\sim}{x}, \underset{\sim}{y})_B = B^{-1}b ,$$

where $(x)_B$ denotes the components associated with basis B.

It suffices to show that $\underset{\sim}{y}_B \geq 0$ for proving $\underset{\sim}{x}$ is a solution to P. Note that

$$\tilde{y}_i - \underset{\sim}{y}_i = B_i^{-1} e \cdot \varepsilon .$$

Since

$$|B_i^{-1} \cdot e| \leq n,$$

we have

$$|\tilde{y}_i - \underset{\sim}{y}_i| \leq n \cdot \varepsilon = \frac{1}{2q} .$$

On the other hand if $\underset{\sim}{y}_i < 0$, then $\underset{\sim}{y}_i \le -\frac{1}{q}$. But we know that $\tilde{y}_i \ge 0$. If $\underset{\sim}{y}_i < 0$ we have a contradiction.

$$\therefore \quad \underset{\sim}{y}_i \ge 0.$$

Hence $\underset{\sim}{x}$ is a solution to $Ax \le b$.

<u>Complexity of a direct ellipsoidal algorithm</u>

Consider the linear programming problem LP.

<u>Lemma 5</u>: Let $f(B)$ represent the value of the objective function at the feasible basis B. If $f(B_1) \ne f(B_2)$ then $|f(B_1)-f(B_2)| \ge \frac{1}{q}$. □

<u>Theorem 6</u>: The ellipsoidal algorithm combined with VERTEX solves LP in at most $M(\varepsilon)$ iterations, for $\varepsilon = (2||c||n^{3/2} q)^{-1}$.

<u>Proof</u>: Let us note that if we represent LP as

$$\text{LP}': \quad \begin{aligned} &\max f(x) = cx \\ &s(x) \le 0, \end{aligned}$$

then the ellipsoidal algorithm gives a point $\bar{x}$ s.t.

$$|f(\bar{x})-f(x^*)| < \varepsilon$$

and $s(\bar{x})-s(x^*) < \varepsilon$, where x^* solves LP.

Let x be a basic point of $P_\varepsilon = \{x: Ax \le b + e \cdot \varepsilon\}$, and $\underset{\sim}{x}$ the corresponding basic point of P, obtained by algorithm VERTEX using the objective function. Since $\tilde{x} - \underset{\sim}{x} = \varepsilon B^{-1} e$, and

$$|\varepsilon c_B B^{-1} \cdot e| \le \varepsilon||c||n^{3/2} = \varepsilon_2$$

i.e. $\quad f(\underset{\sim}{x}) \ge f(\tilde{x}) - \varepsilon_2$,

then

$$f(\underset{\sim}{x}) \geq f(\tilde{x}) - \varepsilon_2 \geq f(\bar{x}) - \varepsilon_2 \geq f(x^*) - \varepsilon - \varepsilon_2.$$

Thus

$$|f(\underset{\sim}{x}) - f(x^*)| < \varepsilon + \varepsilon_2 ,$$

for

$$\varepsilon = \frac{1}{2||c||n^{3/2}q} , \quad \varepsilon_2 = \frac{1}{2q} , \quad \varepsilon \leq \varepsilon_2 .$$

Thus

$$|f(\underset{\sim}{x}) - f(x^*)| < \frac{1}{q} .$$

But, by Lemma 3, we have $f(\underset{\sim}{x}) = f(x^*)$.

For above ε, d, and $R = n||b||$, we have

$$\begin{aligned} M(\varepsilon) &= 2n^2(\log 2||c||n^{3/2} q + \log||c|| + \log n||b||) \\ &= 2n^2(\log 2 + 2 \log||c|| + \tfrac{5}{2} \log n + \log q + \log ||b||) \\ &= O(n^2(2 \log||c|| + 5 \log n + \log q + \log||b||)) \\ &= O(n^2 \log(nq||c|| \, ||b||)). \end{aligned}$$

4.4 SOME MODIFICATIONS

Since the ellipsoidal algorithm has been known as an algorithm for linear programming, almost all modifications have been suggested for linear programming. We will discuss mainly deep cuts, layer cuts (two sided cuts), and surrogate cuts. They are quite natural and readily available in linear programming because of explicit knowledge of the linear constraints.

Deep cuts

Let $E = (x,B)$ be an ellipsoid, and $ax \leq \beta_1$ is violated at $\bar{x}$, i.e. $a\bar{x} > \beta_1$. Previously we formed the new ellipsoid $E^+ = (x^+,B^+)$ containing $\{x \in E: ax \leq a\bar{x}\}$. Now, we form the new E^+ such that $\{x \in E: ax \leq \beta_1\} \subset E^+$. Let

$$e = \frac{ax - \beta_1}{||Ba||} . \tag{1}$$

If $e > 1$, then the hyperplane $ax = \beta_1$ lies outside of E. Thus, for $e > 1$, E does not contain any feasible point, implying there is no feasible point (provided the initial ellipsoid contains all feasible points).

The formulae for the new ellipsoid are the following:

$$s = B^t a/||B^t a||, \tag{2}$$

$$\tau = \frac{1 + ne}{n + 1} , \tag{3}$$

$$\sigma = \sqrt{\frac{n^2}{n^2-1}(1 - e^2)} , \tag{4}$$

$$\beta = \sqrt{\frac{n-1}{n+1}\,\frac{1-e}{1+e}} , \tag{5}$$

$$x^+ = x - \tau Bs , \tag{6}$$

$$B^+ = \sigma B R_\beta(s) . \tag{7}$$

For $e = 0$, we obtain the usual formulae.

Let us prove the above formula. Let

$$\gamma = \gamma(e), \ \alpha = \alpha(e), \ \beta = \beta(e), \ \sigma = \sigma(e),$$

and $x^+ = x - \gamma s$, and assume the current ellipsoid is the unit sphere $E_o = (0,I)$. Since for x s.t. $x = es + y$, $< s,y > = 0$, $e^2 + ||y||^2 = 1$, we have

$$||R_\alpha(s)x - x^+||^2 = \alpha^2(e),$$

which gives

$$(\alpha(e)e - \gamma(e))^2 + ||y||^2 = \sigma^2(e)$$

for

$$x = s,$$

$$||R_\alpha(s)x - x^*|| = (1 - \gamma(e))\alpha(e) = \sigma(e).$$

combining the last two equations with

$$e^2 + ||y||^2 = 1,$$

we have

$$\alpha^2(e) = \frac{1 + e}{1 - 2\gamma + e},$$

$$\sigma(e) = (1 - \gamma)\ \alpha = (1 - \gamma)\sqrt{\frac{1 + e}{1 - 2\gamma + e}}.$$

Then the volume will be

$$V(\gamma) = \sigma^n \quad \beta = (1 - \gamma)^n \alpha^n \cdot \beta = (1 - \gamma)^n \alpha^{n-1}$$

$$= (1 - \gamma)^n \left(\frac{1 + e}{1 - 2\gamma + e}\right)^{n-1/2}.$$

Then

$$\frac{dV}{d\gamma} = 0$$

implies

$$\gamma = \gamma(e) = \frac{1 + ne}{n + 1},$$

thus verifying the above given equations and the corresponding volume reduction:

$$q(e) = \sigma(e)^n \beta(e) = \left(\frac{n^2}{n^2-1}(1 - e^2)\right)^{n/2} \sqrt{\frac{n-1}{n+1}\frac{1-e}{1+e}}$$

$$= \left(\frac{n^2(1-(e)^2)}{n^2 - 1}\right)^{n-1/2} \frac{n}{n+1}(1 - e). \qquad (8)$$

Layer cuts (parallel cuts)

Suppose we have two parellel inequalities

$$\beta_1 \leq ax \leq \beta_2 \tag{9}$$

and $a\bar{x} < \beta_1$, where $\bar{x}$ is the vertex of current ellipsoid. Then the new ellipsoid containing

$$\{x \in E: \beta_1 \leq ax \leq \beta_2\}$$

can be calculated by the formulas:

$$e_1 = \frac{ax - \beta_1}{||Ba||}, \quad e_2 = \frac{\beta_2 - ax}{||Ba||}, \tag{10}$$

$$\sigma = \sqrt{\frac{n^2}{n^2-1}\left(1 - \frac{e_1^2 + e_2^2 - \rho/n}{2}\right)}, \tag{11}$$

$$\gamma = \sqrt{4(1 - e_1^2)(1 - e_2^2) + n^2(e_1^2 - e_2^2)^2},$$

$$\rho = \frac{e_1 - e_2}{2}\sigma, \tag{13}$$

$$\sigma = \frac{1}{n+1}\left(n + \frac{2}{(e_1 - e_2)^2}(1 - e_1e_2 - \rho/n)\right). \tag{14}$$

See for the derivation of the formulae Shor and Gershovich [S 20], Todd [T6], Gill-Murray et al [G 25], Bland et al [B 24].

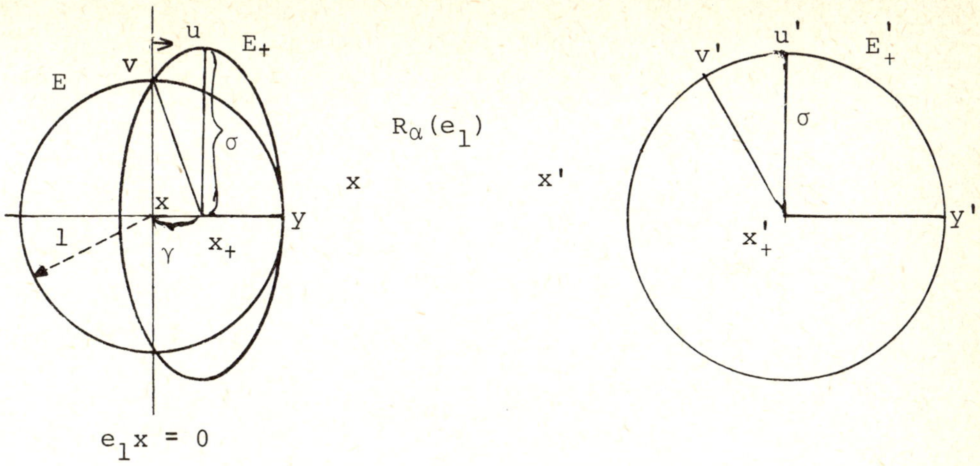

space dilatation operation

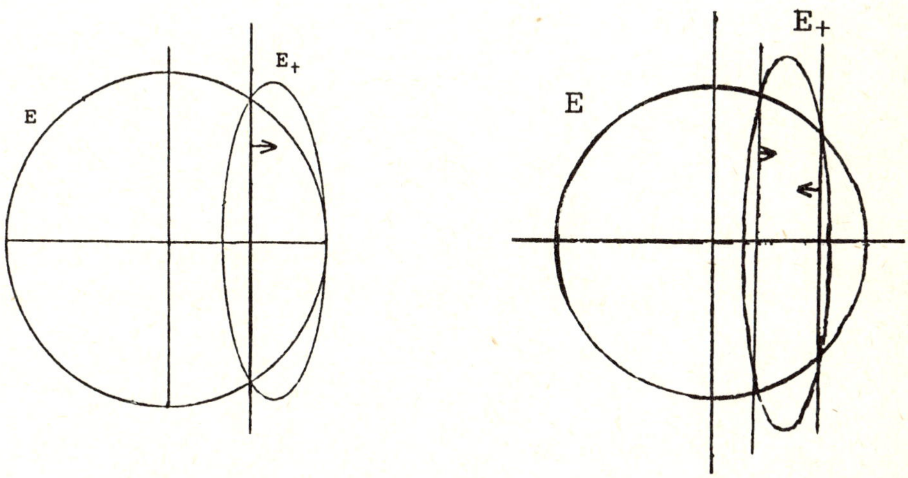

Deep cut

Parallel cut

Fig. 4.1

Surrogate cuts

It is well-known that for $u \geq 0$,

$$(uA)x \leq (ub) \tag{15}$$

is a valid cut for

$$Ax \leq b.$$

Now suppose $||a_1|| = ||a_2|| = 1$ and $< a_1,a_2 > \leq 0$. Let $\bar{x}$ violate both of them. Any attempt to satisfy one of them would probably end with increased violation of the second one. One remedy is to consider both of them at the same time. Let $\theta = < a_1,a_2 >$, $\varepsilon_1 = a_1x - b_1$, $\varepsilon_2 = a_2x - b_2$. Then

$$u = \begin{bmatrix} 1 & \theta \\ \theta & 1 \end{bmatrix}^{-1} \begin{bmatrix} \varepsilon_1 \\ \varepsilon_2 \end{bmatrix} = \frac{1}{1-\theta^2} \begin{bmatrix} \varepsilon_1 - \theta\varepsilon_2 \\ \varepsilon_2 - \theta\varepsilon_1 \end{bmatrix} > 0 \tag{16}$$

provided $\theta \in (-1,0]$, $\varepsilon_1 \geq 0$, $\varepsilon_2 \geq 0$ thus gives the coefficients of the surrogate constraints.

Consider a set of hyperplanes $\{a_i,\ i \in I\}$. If $i,j \neq I$ $< a_i,a_j > \leq 0$, $i \neq j$, we can obtain a surrogate constraint by solving the following quadratic programming problem

$$\max_{u \in 0} u^t(A^t\,\bar{x} - b)/(u^tA^tB^tBAu)^{1/2} \tag{17}$$

where $(\bar{x},B)$ represents the current ellipsoid, and A represents $\{a_i:\ i \in I\}$ (see Goldfarb-Todd [G 40]). It can be show that if

$$\bar{u} = (\bar{A}^tB^tB\bar{A})^{-1}(\bar{A}\bar{x} - \bar{b}) \geq 0, \tag{18}$$

then the surrogate cut $\bar{u}^t\bar{A}x \leq \bar{u}^t\,\bar{b}$ is the deepest cut for the submatrix $\bar{A}$. A sufficient condition for this is $\bar{A}^tB^tB\,\bar{A}$ has non-positive off diagonal elements. In other words, hyperplanes must be obtuse in the metric defined by the current ellipsoid (see Goldfarb-Todd [G40], Todd [T6], Bland et al [B24]). It is clear that the search for a maximal subset of pyperplanes satisfying the above conditions, hence a maximum subset is very difficult, since it is an instance of a NP-complete problem, namely vertex packing problem.

Equality constraints

If among the defining linear system we have

$$A_1 x = b_1,$$

it is necessary to handle these directly. This is so, because writing them as 2 sets of inequalities and applying ellipsoidal algorithm will result in many iterations to satisfy the equality system. Thus the waste of so many iterations can be prevented by elimination of some of the variables, and thus working in a smaller dimension.

If A_1 is $m \times n$ and $r(A) = m$, then one can eliminate m of the variables and thus work with $n - m$ variables. As a result some simple constraints of the form $x \geq 0$, will become more complex. One can choose x_o so that $Ax_o = b_1$, and keep $x^{i+1} - x^i$ in the nullspace of A_1. To achieve the elimination in a numerically stable way one can work with QR decomposition, see, e.g. [S 39].

Let

$$A_1^T = [Q_1 \mid Q_2] \begin{bmatrix} R \\ 0 \end{bmatrix} = Q \begin{bmatrix} R \\ 0 \end{bmatrix}$$

where Q is an orthonormal matrix and Q_1 and Q_2 have m and $n - m$ columns, respectively. The columns of Q_2 and Q_1 span the linear space $\{x: A_1 x = 0\}$, and its orthogonal complement. Let us define y by

$$x = Qy = Q_1 y_1 + Q_2 y_2 .$$

Then clearly $A_1x = b_1$ is equivalent to

$$b_1 = A_1x = A_1Qy = A_1(Q_1y_1 + Q_2y_2)$$
$$= A_1Q_1y_1 = R^T y_1 .$$

Thus, the constraint $A_1x = b_1$ is eliminated, and new variables are y_2 in R^{n-m} . The linear form ax will become

$$ax = aQy = a(Q_1y_1 + Q_2y_2) = (aQ_2)y_2 + (aQ_1y_1)$$

for a fixed y_1. Similarly the constraints $x \geq 0$ will become

$$Q_1y_1 + Q_2y_2 \geq 0$$

or

$$Q_2y_2 \geq -Q_1y_1 .$$

Note that since $R^T y_1 = b_1$, $y_1 = (R^T)^{-1} b_1$, i.e. y_1 is a known quantity.

Another approach is to use projection. It is less efficient. Let x^o be such that

$$A_1x^o = b_1.$$

Let

$$P: R^n \rightarrow R^n$$

such that

$$A_1P = 0.$$

P is the orthogonal projection operator on the nullspace $A_1x = 0$. Since every point of the ellipsoid (x^o,I) is of the form

$$x = x_o + y, \quad ||y|| \leq 1,$$

the ellipsoid (x^o,P) will be in the affine subspace $H_1:\{x: A_1x = b_1\}$. Thus in this way, one can forget about the constraints $A_1x = b_1$, and for each ellipsoid (x,B), the ellipsoid $(x,PB) \subset H_1$. In this case one must work with n-variables.

Equivalently one can start with (x^o,I) as the initial ellipsoid. At each iteration, when one of the equalities is used as a cut, one can perform a slightly different operation; satisfy the constraint as equality, and keep it as equality by means of projection or a similar device. Satisfaction as an equality can be done using a non-orthogonal projection defined by the current ellipsoid.

A dimensional reduction

For the current ellipsoid (x,B), e of (1) represents the eccentrecity of the hyperplane $ax = b$ with regard to center of the ellipsoid in the normalized metric defined by the ellipsoid, i.e. when ellipsoid seen as a unit ball in a suitable space. As indicated previously, if $e > 1$ then whole unit ball lies on the infeasible side of hyperplane, whence the polyhedron is empty. If $e < -1$, then the unit ball, or the current ellipsoid, lies on the feasible side of the hyperplane. Hence it is redundant, and thus can be discarded. If one is solving LP via primal-dual formulation as in Gács-Lovász [G13], then corresponding dual variable must be satisfied as zero in any optimal solution by the complementary slackness of linear programming. In this case, this variable must be eliminated, resulting a dimensional reduction. See Jones and Marvil [J5] for details.

Stability

Early researchers soon realized that the ellipsoidal algorithm was not "stable". When they implemented Gács-Lovász version, term under the square root sign which is the norm in the metric corresponding to the current ellipsoid was negative after some iterations. This phenomena was interpreted as instability. Korte-Schrader [K19] analyzed the eigenvalue structure and observed that condition number of the symmetric positive definite metric representing the current ellipsoid is not well-behaved. They interpreted this as "inherent instability" of the ellipsoid method. A few comments are in order. In the first place, Gács-Lovász version is not the ellipsoidal algorithm. If one implements any of the Hachiyan, Shor or Judin-Nemirovskii versions, this problem can not occur. Secondly, a numerical analysist will tell at the first sight that implementation is not numerically stable. Later, scientific community rediscovered Shor, or Hachiyan versions. Obviously, one can use other numerically stable factorizations of H_k, or $B_k^T B_k$, as suggested by several authors, e.g. Goldfarb-Todd [G40], Gill et al [G25]. In fact Gill et al [G25] showed that the ellipsoid algorithm is "super stable", i.e. even H_k becomes singular, one can still implement the algorithm in a numerically stable way.

Thus numerical stability can be achieved quite simply by using well-developed numerical linear algebra methods.

One should note that in case of convex programming, Ecker and Kupferschmid [E2] reports that they did not have any stability problem with H_k.

4.5 NUMERICAL EXPERIMENTS

After the presentation of Gács-Lovász paper at the Mathematical Programming Symposium in Montréal, in August 1979, many researchers implemented mostly Gács-Lovász version of the ellipsoidal algorithm, though later some implemented different versions of the algorithm. Initially, stability was a major problem. As explained previously, that was due to poor implementation of the algorithm. Now, numerically stable implementations of the algorithm are well-known.

Most of the experiments showed poor convergence. Now, it is a common belief that with the current implementation the ellipsoidal algorithm is not competitive with the simplex method for the type of problems that are encountered in practice.

If one considers the original Shor or Hachiyan algorithm for convex minimization problem or equivalently linear inequality system, one is only guaranteed a geometric convergence with ratio roughly $1- 1/2n^2$. Deep cuts and other improvements do not change this ratio dramatically. As shown by Bland et al [B 24], deep cut only decreases the total number of iterations roughly by a factor of 5, under some assumptions. Since the algorithm tries to reduce the volume of the localization region of solutions, when the required region has essentially zero volume, it will require the maximum allowable number of iterations.

In response to numerous inquires about the ellipsoidal algorithm. Dantzig [D 8] compared then conjuctured expected behaviour of the simplex method with the worst case behaviour of the ellipsoid algorithm. Later McCall elaborated this and similar point. Obviously the simplex method is much better.

Even though numerous researchers experimented with the ellipsoidal algorithm on variety of problems ranging from problems with 3-5 variables to several hundred variables, for randomly generated ones to some banchmark problems very few have reported in the open literature, possibly due to failures. Some of them are, Wang [W 2], Pickel [P 9], Halfin [H 6],

Krol and Mirvan [K 24], Kiehm and Vincke [K16]. The most comprehensive experiment is reported by Gill et al [G25]. They tried various strategies. Their main purpose was to find a feasible point of the perturbed system that would allow the identification of defining hyperplanes of the optimal solution. Result of their experiment was negative. Part of the difficulty comes from the nature of the problem. In case of degeneracy or near degeneracy, it is difficult to identify those hyperplanes even in the simplex method; you have to perform degenerate pivots.

One feature of the ellipsoidal algorithm reported in Gill et al [G25] and Ecker and Kupferschmid [E2] in the context of convex programming is that the algorithm behaves rather well at the beginning, and very stubborn towards to the end. Consequently if the volume of the region in which a point is sought is large then the algorithm is most likely to converge fast. However, when the volume is rather nil, it will require full power of the estimated bound.

Let us recall that the upper bound $6n^2L$ has several components. If one wants to find a ε-feasible (optimal) solution, then the required number of iterations are

$$2n^2 \lceil \log R + \log d + \log \frac{1}{\varepsilon} \rceil ,$$

where R is the radius of the initial ellipsoid, d an upper bound on the subgradients, and ε is the required precision. For practical problems R, d can be estimated to be quite low with regard to L. Validity of the algorithm requires an ε of the order 2^{-L} which makes the upper bound large. For a reasonable ε, the above bound is substantially low. Problem is that one can not do much with such an approximate solution. If it were possible to use such an approximate solution, then the ellipsoidal algorithm could be a practical algorithm.

The ellipsoidal algorithm is an outgrow of relaxation and subgradient methods. It is well-accepted that, at least for sparse large systems, relaxation and subgradient methods are very effective. It is also worth reminding the reader that the algorithm is originally designed for convex, non-smooth problems. It is rather surprising that the practical behaviour of the algorithm so discouraging. One explanation would lie in the fact that in the case of linear programming, the algorithm uses very little of the underlying linear structure; only use is made the characterization of nonsolvability and hence in choice of ε.

Contrary to western literature, forerunners of the ellipsoidal algorithm, namely the same algorithm with different parameters [S19, S21, M18], [I 4] reported or claimed to be very practicle, see also Goffin [G30]. Essentially, same type of algorithms with different parameters are proposed for higher dimensional problem: Shor and Gershovich [S34], Judin-Nemirovskii [J10, N 3]. One possible explanation lies in the fact that in Soviet Union and western countries the level of hardware and software development are in different stages.

Ellipsoidal Algorithm and Integer Programming

Later in Chapter 8, we will study the consequences of the ellipsoid method in combinatorial programming, hence in integer programming. Here our aim is to report on some efforts to combine ellipsoidal algorithm with various integer programming methods.

Since solution of integer programming problem with ellipsoidal algorithm would imply $NP = P$, efforts are directed to hybrid algorithms. That is ellipsoidal algorithm is used in finding good initial or partial solutions to integer programming problem. Consider a 0-1 integer programming problem. If during an iteration of the algorithm, if one can verify that the projection of the current ellipsoid on a particular coordinate contains only one integer point, in this case a zero or one, then clearly that coordinate is fixed to that particular value in any feasible, hence optimal solution to integer programming problem. This is called ellipsoid initialization. Then attempts have been made to use this partial solution within the context of implicit enumeration. Bijlani reports on his algorithm on subset so randomly generated moderate size problems and some Steiner triple problems. Steiner triples are among the hardest integer programming problems, e.g. see Fulkerson et al [F14].

Bijlani reports that his hybrid algorithm is compatitive with Geoffrian's algorithm.

4.6 A NEW LOOK AT RELAXATION METHODS

Goffin [G37] reanalyzed the relaxation methods with regard to ellipsoidal algorithm. One can see the ellipsoidal algorithm as a variable metric algorithm with certain provisions. Fritz John [J 3] in his landmark paper considered the following problem: given a convex polytope with non-empty interior what is the smallest ellipsoid containing the polytope and largest ellipsoid contained in the polytope. Incidentally, in solving this problem he obtained his famous optimality conditions for nonlinear programming. John showed that these two ellipsoids are homothetic. So there is essentially one ellipsoid. If one takes the nonsingular matrix T defining the ellipsoid, transform the space by this matrix, then the relaxation method in the new space can be shown to be polynomial. In other words, use a special ellipsoidal algorithm where the ellipsoid is fixed then the Agmon's relaxation algorithm [A 3] will become polynomial. Goffin showed this result by calculating a condition number of the linear inequality system. One cause of small convergence rate of the relaxation method is the angle between two hyperplane may be too small, hence resulting a sequence of alternating iterates on the two hyperplanes. The above transformation in some sense "opens" this angle in optimal way, thus allowing polynomial convergence.
The transformation obtained by Fritz John ellipsoid in some sense defines an "inverse Hessian" for the linear inequality system. Moreover this inverse hessian can be built conceptially from rank one matrices.

Difficulty lie in the determination of this transformation matrix. In a subsequent paper, Goffin [G38] showed that the ellipsoid method applied with deep cuts is an implementable version of the conceptual algorithm defined through Fritz John matrix.

4.7 AN EXACT RATIONAL ARITHMETIC ALGORITHM

Ursic [U 1] developed an algorithm to solve an equality system with nonnegativity restrictions with exact rational arithmetic. We will give a summary of his algorithm.

Given the following system

$$Ax = b, \quad x \geq 0 \tag{1}$$

where $A \in Z^{m \times n}$, $b \in Z^m$. The basic idea of the algorithm is to handle the constraint $Ax = b$ implicitly. That is try to keep the center of the ellipsoid on the solution space

$$H = \{x: Ax = b\}. \tag{2}$$

When the center of the ellipsoid $\bar{x} \notin H$, project $\bar{x}$ on H. He uses a non-orthogonal projection in order to take care of the shape of the ellipsoid. The basic idea is similar the one we used in Chapter 6.

During the algorithm one needs to take square roots in the computation of β, σ, and s.

Let us define

$$E = \sigma B. \tag{3}$$

The implicit handling and thus elimination of equality constraints leaves one with the simple nonnegativity constraints. In other words, this allows one to use simple cuts of the form

$$e_j x = x_j \geq 0. \tag{4}$$

Since, for

$$g = e_j \, , \quad R_\beta(g)g = \beta g, \tag{5}$$

and by initialization B = I or any diagonal matrix in that matter, one can show by induction that

$$s = B^t g/||B^t g|| \, , \tag{6}$$

$$s = g \, . \tag{7}$$

In other words, under the above hypothesis, the ellipsoid axis always will be parallel to coordinate axis, or equivalently matrix E will be diagonal. Thus this construction allows the elimination of square root in the calculation of in formula (6).

The update formula for E is

$$E^+ = E\,\sigma R_\beta(g) \, . \tag{8}$$

But, for $g = e_j$, $\sigma R_\beta(g)$ is a diagonal matrix:

$$(\sigma R_\beta(e_j))_{ii} = \begin{cases} \sigma\beta & i = j \\ \sigma & \text{otherwise} \end{cases} \tag{9}$$

for $\sigma = \sqrt{\dfrac{n^2}{n^2-1}}$, $\beta = \sqrt{\dfrac{n-1}{n+1}}$. Notice that $\sigma\beta = \dfrac{n}{n+1}$. This will eliminate the square root term due to β. For σ Ursic replaces σ with an upper bound. Notice that

$$\sigma = \sqrt{1 + \frac{1}{n^2-1}} = 1 + \frac{1}{2(n^2-1)} - \frac{1}{8}\left(\frac{1}{1+\theta}\right)^{3/2}\left(\frac{1}{n^2-1}\right) \tag{10}$$

for some $\theta \in (0, \dfrac{1}{n^2-1})$ by Taylor's theorem. Thus, one has

$$\sigma \le 1 + \frac{1}{2(n^2-1)} = \frac{2n^2-1}{2(n^2-1)} = \sigma' .$$

By replacing σ by σ' , one avoids the last square root term. One has to verify that using σ' instead of σ would not change the volume reduction ratio too much. But, that is easy: since

$$\sigma'^{n-1}(\sigma\beta) = (1 + \frac{1}{2(n^2-1)})^{n-1} \frac{n}{n+1} \le e^{\frac{1}{2(n+1)}} \cdot e^{-\frac{1}{n+1}}$$
$$= e^{-\frac{1}{2(n+1)}} , \qquad (11)$$

we have the usual guaranteed volume reduction.

This completes the validity of the algorithm. To obtain the upper bound is another matter. Gács-Lovász type arguments do not work, since solutions lie in a subspace. However, Judin-Nemirovskii and Hachiyan type proofs (see Chapter 3) remain valid. Ursic gives another argument, by which complexity is increased by n times. Recall that the axis of the ellipsoid remain parallel to original coordinate axis. Thus, at each iteration one can determine the projection of the ellipsoid on any coordinate axis. When the length of this interval is less than $1/2q^2$, where q is the a priori upper bound on the determinants of the system, or equivalently largest numerator/denumerator in the basic points of the linear system, it follows from the theory of continued fractions (see Chapter 3, Section 3.8) there is only one rational points which can be part any basic point. One can show easily that the complexity of this is the same as the complexity of original ellipsoidal algorithm. Thus one fixes a coordinate or shows no solution exists and hence repeating n times finds a solution.

We would like to point out that the above algorithm is <u>not</u> polynomial independent number of constraints modula a feasibility oracle. One might recall that the original Hachiyan version is not either. But, that could be changed by assuming an oracle to check the feasibility and by putting an a priori upper bound on the coefficients of the linear system as Ursic does.

Let us consider the linear inequality system

$$Ax \leq b \ , \tag{12}$$

where A is $m \times n$. Even if we neglect the increase due to changing x to non-negative x, i.e. writing $x = x^+ - x^-$, $x^+ \geq 0$, $x^- \geq 0$, we have to add m slack variables in order to write (12) as

$$\begin{aligned} Ax &= b \\ x &\geq 0 \ . \end{aligned} \tag{13}$$

That is n increases to n + m. This increases the complexity considerable. Furthermore, at each iteration, in terms of original system, one has to solve, $(2m + n) \times 2(m + n)$ system in exact arithmetic.

Instead one can work in the affine subspace H, which has the same dimension with the original one, such an approach is used in Chapter 6, and also described in Goldfarb-Todd [G40].

4.8 HALFIN'S SPHERE METHOD

Jones and Marwil [J4], and independently Halfin [H6] suggested a 'sphere' method instead of ellipsoid method. We will analyze Halfin's method. Method is hardly new. Apparently Halfin, is unaware of Shor's 1970 [S13] and 1977 papers [S18-19].

Let us recall Shor's space dilatation operator. For

$$||s|| = 1, \quad \alpha > 0,$$

$$R_\alpha(s) = I + (\alpha - 1)ss^T. \tag{1}$$

(See Chapter 3, Section 3.7 , for details.) Shor at iteration k assigns another copy of R^n, so that when one applies $R_\alpha(s)$'s sequentially arrives a sphere of radius $R\sigma^k$ centered at x^k, as the image of the current ellipsoid (x^k, B_k). Shor, then takes the necessary subgradient operation, form the new ellipsoid, but at the same time carries back center of the ellipsoid into original space. Halfin works in the transformed space and he transforms the linear system in terms of new variables, and thus solves the problem in the sequence of transformed spaces. He does not calculate the center of the ellipsoid explicitly. Instead he updates the residues, i.e. $a_i x^k - b_i$. When the algorithm stops, i.e. when the residues are all non-positive, he has to resort some linear algebra tricks. If for example, the algorithm stops at an interior point, then the sphere method is in trouble. He assumes the strict inequality system, but works with (regular) inequality system.

As noted by Bland et al [B24] the relation with sphere method and other approaches to ellipsoid method is similar to tableau form simplex and revised simplex method. Clearly sphere method implementation of the ellipsoid method takes more work than the usual implementation of in ellipsoid method.

There is an attempt to analyze the behaviour of the underlying polyhedron in the sequence of transformed spaces. Let us remark that Halfin's space dilatation operator is multiplied by σ. For the following analysis, it has no effect, so we will use our notation.

Theorem: Let $||a|| = ||s|| = 1$, $0 < \beta < 1$, then

$$\frac{|< R_\beta(s)s,\ R_\beta(s)a >|}{||R_\beta(s)s||\ ||R_\beta(s)a||} \leq |< a,s >| \tag{2}$$

Proof: Recall that

$$R_\beta(s)s = \beta s, \quad R_\beta^T(s) = R_\beta(s). \tag{3}$$

Then

$$< R_\beta(s)s,\ \ R_\beta(s)a > = <\beta\, s,\ R_\beta(s)a > = \beta < a,\ R_\beta(s)s >$$

$$= \beta^2 < a,s > . \qquad (4)$$

Thus (2) becomes

$$\frac{\beta^2 |<a,s>|}{\beta ||R_\beta(s)a||} \le |<a,s>|. \qquad (5)$$

Recall that

$$||R_\beta(s)a|| = \sqrt{||a||^2 + (\beta^2-1) < s,a >^2}$$

$$= \sqrt{1 + (\beta^2-1) <s,a >^2} \quad . \qquad (6)$$

If $<a,s>=0$, or $<a,s> = \pm 1$, then (5) or (2) holds as equality. Otherwise, we have $0 < | <a,s>| < 1$. Now (5) is equivalent to

$$||R_\beta(s)a|| \ge \beta \ . \qquad (7)$$

Let

$$a = \delta s + \bar{s},\quad \delta \in R \quad , \quad \bar{s} \in R^n,\quad \delta^2 + ||\bar{s}||^2 = 1,$$

$$<s,\bar{s}> = 0. \qquad (8)$$

By assumption $\bar{s} \ne 0$. Then

$$R_\beta(s)a = \beta\, \delta\, s + \bar{s}. \qquad (9)$$

$$||R_\beta(s)a||^2 = \beta^2\delta^2 + ||\bar{s}||^2 > \beta^2\delta^2 + \beta^2||\bar{s}||^2 = \beta^2 \qquad (10)$$

□

This states that each hyperplane becomes "more orthogonal", to the hyperplane used in the cut.

Let us estimates how these quantities changes. Let

$$\theta = \frac{\beta}{\sqrt{1 + (\beta^2 - 1)\langle s,a\rangle^2}}$$

since

$$\beta = \sqrt{n-1/n+1} \ ,$$

$$\theta^2 = \frac{\beta^2}{1 + (\beta^2 - 1)\langle s,a\rangle^2} = \frac{n-1}{n+1-2\langle s,a\rangle^2}$$

$$\theta \approx 1 - \frac{1 + \langle s,a\rangle^2}{n+1-2\langle s,a\rangle^2} \ .$$

Finally, we would like to comment on a remark by Halfin. He sees a similarity between simplex pivot and space dilatation operation. Unfortunately, similarity is on the surface; one is a vector addition and the other is a space dilatation operation.

4.9 AN ACCELERATED DIRECT ELLIPSOIDAL ALGORITHM

We will present an acceleration of the direct algorithm which has been discussed above. The acceleration that we suggest is the following: when the current point x_k is feasible, make a virtual move to a basic point, $\tilde{x}_k$ of the underlying polyhedron, while improving the objective function. At $\tilde{x}_k$ check whether LP is optimal, or unbounded. If yes, then we stop. Otherwise, we use $f(\tilde{x}_k)$ in calculating the deep cuts for the objective function until a better value of the objective function is obtained. In the virtual move from x_k to $\tilde{x}_k$, we suggest to use the ellipsoidal gradient method, i.e. $s_k = B_k B_k^t c$, if possible. Otherwise we can use $s = c$. We first hit one of the boundaries of the feasible region using the direction s, say at $\bar{x}$. Then in at most (n-1) iterations of the simplex method or the projection method, one reaches $\tilde{x}_k$. The move from $\bar{x}$ to $\tilde{x}_k$ is discussed previously as the algorithm VERTEX, can be performed in polynomial time, in fact in $O(n^3)$ elementary operations.

The algorithm solves some well-known hard LP problems, at the worst linearly; namely Klee-Minty [K14], Clausen [C13], Avis-Chvátal [A21], Edmonds Shortest Path [C37], Goldfarb-Sit [G39].

The basic idea of the LP algorithm can be adopted to linear complementarity problems (LCP) with certain properties. When the objective function associated with LCP is convex, the obvious extension of the above algorithm solves the exponential time examples of Murty [M29], and Fathi [F5].

In LP and LCP, when a better objective function value is found, it is added to system implicitly. With proper implementation the algorithm stays polynomial.

Algorithm

We will represent an ellipsoid by (x,B), where x is the center, and B is a non-singular matrix. The ellipsoid (x,B) is $\{x + By: ||y|| \le 1\} = \{y: (y-x)^t (B^t B)^{-1} (y-x) \le 1\}$.

We will consider the problem

$$\text{LP} \quad \begin{array}{ll} \max & cx \\ \text{s.t.} & Ax \le b \\ & 0 \le x \le d. \end{array} \qquad \begin{array}{r} (1) \\ (2) \\ (3) \end{array}$$

The version of the ellipsoidal algorithm used to solve

$$\text{P} \quad \begin{array}{ll} \text{find } x & \\ \text{s.t. } Ax \le b & \end{array}$$

is the usual one, say, similar to one given by Philip Wolfe [W 8].

0. $k = 0$, $(0, B_o)$ is the initial ellipsoid

1. Is $a_i x \le b_i$, $\forall i$? If yes, stop.

   Otherwise let $a_i x > b_i$:

$$e = (a_i x - b_i)/||Ba_i|| \ ,$$

$$x_+ = x - \frac{1 + ne}{n+1} \frac{BB^t a_i}{||B^t a_i||} \ ,$$

$$B^+ = r(e) B R_{\beta(e)}(s) \ ,$$

$$\text{where} \quad s = \frac{B^t a_i}{||B^t a_i||} \ , \ R_\beta(s) = I + (\beta - 1) s s^t, \ \text{for } ||s|| = 1,$$

$$r(e) = \sqrt{n^2(1-e^2)/n^2-1} \ , \qquad \beta(e) = \sqrt{\frac{n-1}{n+1} \frac{(1-e)}{(1+e)}} \ .$$

Constraints (3) are handled similarly.

When x is feasible for (2), and (3):

Let $u = Bs$, and if u is a feasible direction determine $t(u)$ s.t. $\tilde{x} = x + t(u)u$ hits one of the boundaries of (2) and (3).

Otherwise consider $u = c$, as virtual feasible direction. If both of these fail, let $\tilde{x} = x$, go to algorithm VERTEX.

Thus the algorithm suggested solves a sequence of problems: find x s.t.

$$P_\nu: \quad \begin{aligned} Ax &\le b \\ 0 \le x &\le d \\ -cx &\le -f_\nu \ , \end{aligned}$$

where $f_o = -\infty$.

If x_ν is a solution to P_ν, and $\tilde{x}_\nu$ is the corresponding basic point given by VERTEX, then

$$f_{\nu+1} = c\tilde{x}_\nu \ .$$

If $\tilde{x}_\nu$ is not a solution (i.e. optimal or unbounded), then we move to $P_{\nu+1}$. In that case the starting point for $P_{\nu+1}$ is x_ν .

In order to obtain the polynomial bound one must be very careful in implementing the ellipsoidal algorithm to solve P_ν. An objective function cut is used during the algorithm only if the current point is ε-feasible (or feasible). Otherwise we do not have the polynomial bound.

This algorithm can be seen as a hybrid algorithm. Various ways of combining the ellipsoidal algorithm with a simplex method are suggested by several researchers (see Bland et al [B 24] and Schrader [S 5] for a general survey). When one reaches a vertex of the LP, a polynomial number of times application of the simplex pivot retains the polynomiality of the total algorithm. One basic difference between our algorithm and others, in the theoretical level, is that we allow ε-feasibility, whereas others do not.

4.10 KLEE AND MINTY PROBLEM

Klee and Minty [K14], exhibited a set of problems for each n, such that the simplex method will take exponential time. The search direction they considered was Dantzig rule, or largest reduced cost, but they clearly state that a similar polytope can be devised for different schemes.

Let P_n be:

$$\begin{array}{lll} & \max & e_n x \\ \text{s.t.} & -x_1 & \le 0, \\ & x_1 & \le 1, \\ & \varepsilon x_1 - x_2 & \le 0, \\ & \varepsilon x_2 + x_2 & \le 1, \\ & \cdots\cdots\cdots & \\ & \varepsilon x_{n-1} - x_n & \le 0, \\ & \varepsilon x_{n-1} + x_n & \le 1, \\ & x_1, \ldots, x_2 \ge 0, & \end{array}$$

where $\varepsilon \in (0,1/2)$, and e_n is the n'th unit vector.

Let us note that we can obtain the bounds:

$$0 \le x_i \le 1 \ , \quad \forall i. \tag{1}$$

P_n has non-integer coefficients. By choosing ε as a rational number, and multiplying by its denominator we can transform P_n to one with integer coefficients.

However, instead we will apply the algorithm described in Section 9.

Let us represent P_n as

$$P_n^1 \qquad \begin{aligned} \max \quad & cx = e_n x \\ & Ax \le b \\ & 0 \le x \le 1, \\ & x \in R^n, \ A \in \mathcal{R}^{2n\times n}, \ b \in \mathcal{R}^{2n} . \end{aligned}$$

In addition we will show 2 variants of the algorithm:

(i) Since $0 \le x_i \le 1$, $\forall i$, then $||x^+|| \le \sqrt{n}$. Choosing $B = \sqrt{n}\, I_n$, the regular ellipsoid step will give

$$x^1 = \frac{\sqrt{n}}{n+1} e_n \ .$$

Let

$$\alpha = \frac{\sqrt{n}}{n+1} \ .$$

Otherwise $\alpha < 1$.

Clearly x^1 is feasible. Let us apply VERTEX to x^1. Let us imbed this system into:

$$[A,I](x,y) = b,$$
$$e_n x \ge \alpha.$$

Rewriting as $A_1x = b$, $e_nx \geq \alpha$, $e_n, c, x \in R^{2n}$ we have

$$x^1 = (0,\ldots,0,\alpha,0,1,0,1 \ldots \alpha, 1-\alpha),$$

$$c = (0,\ldots,0,1,0,\ldots\ldots\ldots, \quad 0).$$

We take last n columns as the basis. With the usual simplex type argument, we increase x_n. The new solution x, parametrized as a function of x_n, can be represented compactly as

$$x(t) = x^1 + t(0, \quad 0,1 \quad , \quad 1, -1) \tag{2}$$

n'th

$$(0, \quad , \alpha + t,0,1,0,1, \alpha + t, \quad 1-\alpha-t).$$

From the feasibility conditions, and (1), (2), we have

$$0 \leq \alpha + t \leq 1, \quad 1 - \alpha - t \geq 0 \Rightarrow \alpha + t = 1, \ t = 1 - \alpha,$$

which gives

$$x(1 - \alpha) = (0,0,\ldots,1,0,1,0,1,\ldots,1,0),$$

or in the original variables $x = e_n$, which is optimal. Note that for $B_o = RI_n$ for any $0 < R < n + 1$ the above algorithm works.

(ii) Suppose $R = \sqrt{n}$, $x_o = 0$, $B_o = RI_n$.
Then the (regular) ellipsoidal direction is e_n. If we do a feasible direction step, we end up with $t = 1$, $x^1 = e_n$, which is optimal.

Let us note that if we take $c = \gamma\, e_n$ for $\gamma > 0$, the above arguments hold. In particular if $\gamma = 2^{2^n}$, this version ellipsoidal algorithm solves the Klee-Minty problem instantly.

One may think that the Klee-Minty example is too simple. It is not. The success of the ellipsoidal algorithm lies its non-local character. In the second example we will see the usefulness of selecting an appropriate initial ellipsoid.

4.11 CLAUSEN'S PROBLEM

Jens Clausen presented the problem at the Danish-Polish Operations Research Societies meeting at Rolifhed, Denmark in May 1979, (see [C13]).

$$P_n: \quad \max \sum_{j=1}^{n} c_j x_j \tag{1}$$

$$\text{s.t.} \quad \sum_{j=1}^{n} a_{ij} x_j \le 5^{i-1}, \quad i = 1,\ldots,n , \tag{2}$$

$$x_j \ge 0 ,$$

where, $c_j = (4/5)^{j-1}$, $a_{ii} = 1$,

$$a_{ij} = 2(5/4)^{i-j} \text{ for } i > j, \quad a_{ij} = 0, \quad i < j.$$

First let us observe that $x_i \le 5^{i-1}$ is valid for $\forall i$. Thus we have

$$0 \le x_i \le 5^{i-1} . \tag{3}$$

Clearly $x^o = \underline{0} \in R^n$ is a feasible solution. We will choose B_o using (3) as

$$B_o = \text{diag}(1,5,5^2,\ldots,5^{n-1}) \text{ i.e. } (B_o)_{ii} = 5^{i-1}$$

then the ellipsoidal direction, up to a scale, is

$$u = B_o^t B_o c = (1,20,20^2,\ldots,(20)^{n-1}).$$

Clearly u is a feasible direction. Let us compute $t(u)$ s.t. $x = x^o + t(u)u = t(u)u$ hits one of the boundaries of the polytope defined by (2). Let t_i be such that

$$a_i^t u t_i = 5^{i-1} . \tag{4}$$

Clearly we have,

$$t_i \le \frac{5^{i-1}}{20^{i-1}} = (\tfrac{1}{4})^{i-1} ,$$

and

$$t(u) \le \min_i t_i \le (\tfrac{1}{4})^{n-1} .$$

Also t(u) is determined by $a_n^t u t(u) = 5^{n-1}$. In other words $x^+ = t(u)u$ hits $a_n x = b_n$.

Let us add slack variables. The transformed system is

$$(A,I)\underline{x} = b, \tag{5}$$

where $\underline{x} \in R^{2n}$ corresponding to x^+; we have

$$\underline{x} = (t, 20t, 20^2 t, 20^{n-1}, 1-t, *, *, *, 0)$$

where * is a (positive) number.

Corresponding to the artificial basis, we have the reduced cost coefficients

$$c = (1, 4/5, (4/5)^2, (4/5)^{n-1}, 0, \ldots, 0).$$

The first n variables are all candidates to increase. Suppose we choose the first variable to increase, (its reduced cost is largest and it is the first). Since the n'th slack variable is zero, we can not increase the first variable. Recall we have an artificial basis, the n'th basic variable is zero, and the corresponding coefficient in its column is positive. Hence, we switch the first variable with n'th unit vector in the basis. Thus we have decreased the number of non-zero non-basic variables by one.

After pivoting, the cost row, and the n'th row become:

n'th	1	4/5	$(4/5)^2$	$(4/5)^2$	$\frac{1}{2}(4/5)^{n-1}$	0	0	$\cdots$	$+\frac{1}{2}(4/5)^{n-1}$
c	0	0	0	0	$\frac{1}{2}(4/5)^{n-1}$	0	0	$\cdots$	$-\frac{1}{2}(4/5)^{n-1}$.

Then we choose the n'th variable to increase. It will increase until x_1 becomes zero. Then we switch x_1 and x_n; x_n goes into the basis, x_1 goes out of the basis. This tableau is the initial one, except the cost row and x row which are

c	-1	$-\frac{4}{5}$	$-(\frac{4}{5})^2$	$-(\frac{4}{5})^{n-2}$	0	0	0	0	0	0
x	0	*	*	*	*	*	*	*	*	0 .

Then in the remaining (n-2) iterations each x_i will decrease to zero for $2 \leq i \leq n - 1$. Thus at the n'th pivot we will reach the optimal solution.

Let us note that at the first iteration if we choose any other variable, the next tableau will be identical unless we choose the n'th variable. In that case we gain one pivot.

4.12 EDMONDS' SHORTEST PATH PROBLEM

Jack Edmonds exhibited a class of shortest path problems $P_n (n \geq 2)$ for which the simplex method would take exponential time.

More on this can be found in [C37].

Let P_n be the shortest path problem on the graph G_n.

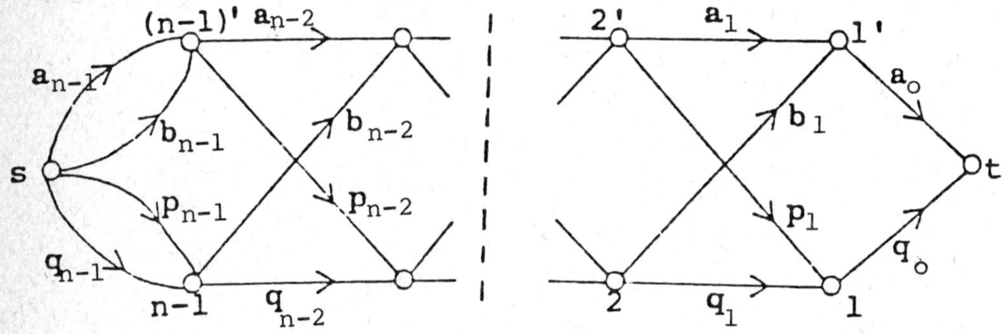

$$|V(G_n)| = 2n \qquad |E(G_n)| = 4(n-1) + 2 = 4n-2$$

$$E = \{a_o, q_o, a_1, b_1, p_1, q_1, \ldots\ldots, a_{n-1}, b_{n-1}, p_{n-1}, q_{n-1}\} .$$

All demands are zero except at s and t, which are 1. Let T_o be the tree $\{a_o, a_1, p_1, \ldots, a_{n-1}, p_{n-1}\}$. Let

$$c(b_i) = c(q_i) = 0, \quad c(a_i) = c(p_i) = 2^i$$

where c(x) denotes the cost of edge x. Clearly $E - T_o$ is the unique optimal tree and the optimal solution is

$$q_i = 1, \quad i = 0, \ldots, n-1.$$

The Shortest Path Problem P_n, like any other flow problem, can be cast as a linear program:

$$\begin{array}{ll} \min & c^T x \\ \text{s.t.} & Ax = b \\ \text{LP} & x \geq 0, \end{array}$$

where $c,\ x \in R^E$, $A \in R^{V\times E}$.

We will transform LP into an equivalent one in less variables, and in inequality form. Obviously using node t, we have

$$a_o + q_o = 1. \tag{0}$$

From node 1,

$$p_1 + q_1 = q_0. \tag{1}$$

From node i,

$$p_i + q_i = b_{i-1} + q_{i-1}. \tag{i}$$

Consider the following cut

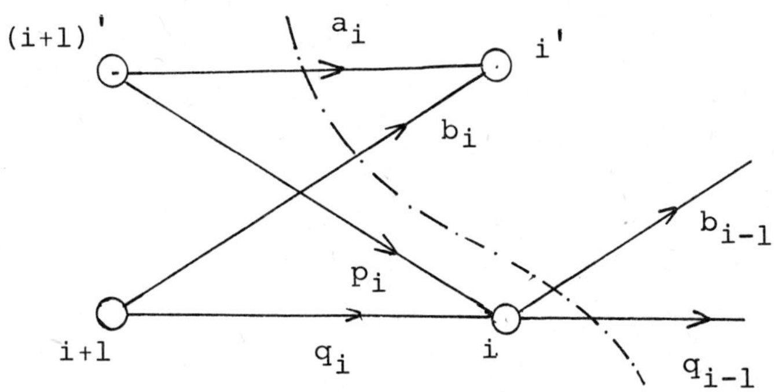

$$a_i + b_i + b_{i-1} + q_{i-1} = 1 \qquad \text{(i')}$$

At node s,

$$a_{n-1} + b_{n-1} + p_{n-1} + q_{n-1} = 1. \qquad \text{(n)}$$

Using the equations (0),(1),(i),(i'),...,(n) we can solve a_i, p_i in terms of b_i, q_i:

$$a_o = 1 - q_o,$$

$$a_1 = 1 - q_o - b_1,$$

$$p_1 = q_o - q_1.$$

For $i \geq 2$

$$a_i = 1 - b_i - b_{i-1} - q_{i-1},$$

$$p_i = b_{i-1} + q_{i-1} - q_i. \qquad (*)$$

Thus we have

$$\min \quad a_o + \sum_{i=1}^{n-1} 2^i(a_i + p_i)$$

LP's: s.t.

$$a_o, q_o \geq 0,$$

$$a_i, b_i, p_i, q_i \geq 0, \quad i = 1,\ldots,n-1.$$

Using (*) the objective function can be written as

$$1 - q_o + \sum_{i=1}^{n-1} 2^i[1 - b_i - q_i]$$

$$= \sum_{i=0}^{n-1} 2^i - [q_o + \sum_{i=1}^{n-1} 2^i(b_i + q_i) \;.$$

Thus LP is equivalent to

$$\max \quad q_o + \Sigma\, 2^i (b_i + q_i)$$

LP": s.t.

$$A_1 x \le d_1,$$

where $A_1 \in R^{m\times n_1}$, $b_1 \in R^m$, $m = 2n-1$, $n_1 = 2n-1$.

A_1 is

	q_o	b_1	q_1	b_2	q_2	b_3	q_3			b_{n-1}	q_{n-1}	d_1
a_o	1											1
a_1	1	1										1
p_1	-1		1									0
a_2		1	1	1								1
p_2		-1	-1		1							0
												⋮
a_{n-1}								1	1	1		1
p_{n-1}								-1	-1		1	0

or explicitly,

$$q_o \le 1,$$

$$q_o + b_1 \le 1,$$

$$-q_o + q_1 \le 0,$$

$$\left.\begin{array}{l} b_i + b_{i-1} + q_{i-1} \le 1, \\ q_i - q_{i-1} - b_{i-1} \le 0, \end{array}\right. \quad (i \ge 2)$$

$$b_i, q_i \ge 0, \ \forall i \quad (b_o = 0).$$

Thus, P_n can be cast as

$$\max \quad c^t x$$

$$LP_n: \qquad A_1 x \le d_1 ,$$

where $c, x \in R^{n_1}$, $A_1 \in R^{m\times n_1}$, $d_1 \in R^m$,

$$m = 2n-1, \quad n_1 = 2n-1.$$

Since $x_i \le 1, \quad i = 1,2,\ldots,n^1$,

$$||x|| \le \sqrt{n_1} .$$

We solved P_n for all n, $3 \le n \le 25$, by computer, with $x_o = 0$, $B_o = \sqrt{n_1}\, I_{n_1}$. In all cases,

the number of ellipsoidal steps was n,
the number of VERTEX (simplex) steps was 2n,

i.e. a total of 3n steps.

We should note that after finding a feasible point, hitting a boundary using a ellipsoidal step was essential. Without that step it was impossible to reach the same complexity.

4.13 AVIS AND CHVATAL PROBLEM

D. Avis and V. Chvátal [A21] exhibited a set of problems, one for each n, such that Bland's rule will solve in exponential time.

Let $0 < \epsilon < 1/2$, and consider

$$\max \sum_{j=1}^{n} \epsilon^{n-j} x_j$$

LP:

$$\text{s.t.} \quad 2 \sum_{j=1}^{n-1} \epsilon^{i-j} x_j + x_i \le 1, \qquad i = 2,\dots,n\ ,$$

$$x_1 \le 1, \quad x_i \ge 0, \quad i = 1,\dots,n\ .$$

Observation 1: $x_i \le 1, \quad \forall i = 1,\dots,n$.

Observation 2: $\underline{x} = \underline{0}$ is feasible.

Observation 3: $x^* = (0,0,\dots,0,1) = y^* = e_n$ form an optimal pair of primal and dual solutions.

Observation 4: $u = c$ is ellipsoidal direction.

Let t_i be such that $a_i(x + t_i c) = b_i$

(assuming LP has the form $\max\{cx | Ax \le b,\ x \ge 0\}$),

$$t_1 = \epsilon^{1-n}\ ,$$

$$c = (\epsilon^{n-1}, \epsilon^{n-2}, \dots, \epsilon, 1),$$

$$t_2 = \frac{\epsilon^{2-n}}{2\epsilon^2 + 1}\ ,$$

$$t_i = \frac{(c_i)^{-1}}{2 \sum_{j=1}^{i-1} \theta^j + 1} = \frac{\epsilon^{i-n}}{2 \sum_{j=1}^{i-1} \theta^j + 1}$$

where $\theta = \epsilon^2$,

and

$$t_n = \frac{1}{2 \sum_{j=1}^{n-1} \theta^j + 1} .$$

Clearly $t_{i+1} \le t_i$. Thus

$$t = t_n = \min_{1 \le i \le n} t_i .$$

Then

$$\bar{x} = \bar{x}(t) = tc = (t\epsilon^{n-1}, t\epsilon^{n-2}, \ldots, t).$$

Let us consider the augmented tableau $(A,I)\bar{x} = b$, where $x \in R^{2n}$, the last n-components of x as slack variables. The basis used in VERTEX is the artificial basis.

	1					1									1
	2ϵ	1					1								1
	$2\epsilon^2$	2ϵ	1					1							1
	$2\epsilon^3$	$2\epsilon^2$	2ϵ	1					1						1
	$2\epsilon^4$	$2\epsilon^3$	$2\epsilon^2$	2ϵ	1					1					1
	....	....	....	....	....	....	....	....	....	....	....	....	....	....	
						1					1				1
							1					1			1
	$2\epsilon^{n-1}$						2ϵ	1					1		1
x	$t\epsilon^{n-1}$		,	$t\epsilon^2$	,		$t\epsilon$,	t	x_{n-1}	x_{n-2}		x_{2n-1}		0	
c	ϵ^{n-1}	ϵ^{n-1}					ϵ	1	0	0		0		0	

Obviously

$$x_{n+1} \ge x_{n+1} \ge \cdots \ge x_{2n-1} \ge x_{2n} = 0,$$

where $x_{n+i} = 1 - a_i\underline{x}$, and $\underline{x}$ is the first n-component of x. Since

$$\max_i c_i = 1,$$

at i = n, we try to increase x_n. But, since $x_{2n} = 0$, we switch x_n, with x_{2n}, pivot on $a_{n,n}$. The new cost row and x row will be

$$\begin{array}{llllll} c & -t\epsilon^{n-1} & , & -t\epsilon^{n-2} & , & -t\epsilon,0,0,0,\ldots,-t \\ x & t\epsilon^{n-1} & , & t\epsilon^{n-2} & , & t\epsilon,t,x_{n+1}, x_{n+2},\ldots,x_{2n}. \end{array} \qquad (\dagger)$$

Let

$$z_i = \epsilon^{n-i} \frac{\left(2\sum_{j=1}^{i-1}\theta^j + 1\right)}{2\sum_{j=1}^{n-1}\theta^j + 1}.$$

Clearly

$$z_1 \le z_2 \le z_i \le z_{i+1} \cdot\cdot \le z_n = 1.$$

We have

$$x_{n+i} = 1 - z_i , \forall i \le n.$$

If we select the minimum c_j, then we choose x_{n-1}. We decrease it to zero. The new x vector will be

$$(t\epsilon^{n-1},\ldots,t\epsilon^2,0,\ t(1 + 2\theta^2),\ x_{n+1},\ x_{n+2},\ x_{2n-1} + t\epsilon,\ 0)$$

Since vector c remains the same, we choose $x_{n-2}, x_{n-3},\ldots,x_1$, in succession, and reduce them to zero. Thus

$$x_i = 0 , 1 \le i < n.$$

Then x_n becomes

$$t(1 + 2\theta^2 + 2\theta^4 + \cdots + 2\theta^{n-1}) \equiv 1.$$

Necessarily

$$x_{n+i} = 1, \quad 1 \le i < n,$$

$$x_{2n} = 0.$$

Suppose we choose x_1 to increase. But, since $x_{2n} = 0$, and x_{2n} is basic

$$x_{2n}(\tau) = x_{2n} - \tau \cdot 2 \cdot \epsilon^{n-1} .$$

For

$$a_1 = (1, 2\epsilon, 2\epsilon^2, \ldots, 2\epsilon^{n-1}),$$

(the first column), we have $\tau = 0$. Thus we switch x_1 with x_{2n}, and we pivot on $a_{n,1}$. The cost row becomes,

$$\underset{}{(0,0,0,\ldots,}\underset{n'th}{1/2}\underset{}{,0,0,\ldots,}\underset{2n'th}{-1/2}),$$

and n'th column becomes,

$$(-1/2\epsilon^{1-n}, \ldots, -\epsilon^{-2}, -\epsilon^{-1}, 1/2\epsilon^{1-n}).$$

Hence we increase x_n. Since the only positive element in the n'th column is $a_{n,n} = 1/2\epsilon^{1-n}$, we have,

$$x^1(\tau) = x - \tau\, a^n ,$$

$$x_1^1(\tau) = x_1 - \tau/2\epsilon^{1-n} = t\epsilon^{n-1} - \tau/2\epsilon^{1-n} .$$

We choose τ such that $x_1(\tau) = 0$.

$$\therefore \tau = 2t\, \theta^{n-1} ,$$

$$x_n = t(1 + 2\theta^{n-1}).$$

Then x_n enters the basis, x_1 leaves the basis. After pivoting on $a_{n,n}$, we obtain the same initial tableau except for the cost row and x-row.

The cost row and x-row become

$$\begin{array}{lllllll} c & -t\epsilon^{n-1}, & -t\epsilon^{n-2}, & -t\epsilon, 0,0,0, \ldots\ldots\ldots & & 0, & -t \\ x & 0, & t\epsilon^{n-2}, & t\epsilon, & t(1 + 2\theta^{n-1}), *, *, \ldots, *, *, & & 0 \end{array}$$

We have returned to the previous case (†). That is, we drive each x_i, $1 < i < n$, to zero in succession.

Thus in n steps we search the unique optimal solution.

4.14 GOLDFARB AND SIT'S PROBLEM

D. Goldfarb and Y. Sit [G 39] give a set of problems which would take 2^n steps in the all variable steepest edge simplex method.

$$\max \quad \Psi_n(x) = \sum_{i=1}^{n} \beta^{i-1} x_i \tag{1}$$

s.t.

LP: $$0 \le x_1 \le 1 \tag{2}$$

$$\beta x_{j-1} \le x_j \le \delta_j - \beta x_{j-1}, \quad j = 2,\ldots,n, \tag{3}$$

where

$$\delta_i = \delta^{i-1} = (\theta\beta)^{i-1}, \quad \theta > 2, \quad \beta \ge 2. \tag{4}$$

Clearly we have the relations

$$0 \le x_i \le \delta_i, \quad i = 1,\ldots,n. \tag{5}$$

Let us rewrite (3) as

$$\beta x_{j-1} + x_j \le \delta_j , \qquad j = 2,\dots,n \tag{6.a}$$

$$\beta x_{j-1} - x_j \le 0 . \tag{6.b}$$

Let A be

$$\begin{array}{cccccc|c}
 & & & & & & b \\
1 & & & & & & 1 \\
\beta & 1 & & & & & \delta \\
 & \beta & 1 & & & & \delta^2 \\
\dots & \dots & \dots & \dots & \dots & \dots & \\
 & & & \beta & 1 & & \delta^{n-1} \\
\beta & -1 & & & & & 0 \\
 & \beta & -1 & & & & 0 \\
 & & & \beta & -1 & & 0 \\
 & & & & & & 0
\end{array} \tag{7}$$

We can write LP compactly as

$$\max \quad \Psi_n(x) = cx \tag{8}$$

$$\text{s.t.} \quad Ax \le b,$$

where $A \in R^{2n-1\times n}$, $b \in R^{2n-1}$ as described.

Since $0 \le x_i \le \delta_i$, we can take as initial ellipsoid (x_o, B_o), $B_o = r \operatorname{diag}(1,\delta,\delta^2,\dots,\delta^{n-1})$ for $1 \le r \le \sqrt{2}$. $x_o = \underline{0}$ is feasible, $u = B_o B_o^T c$ is the ellipsoidal direction. Note that

$$u_i = \delta_i^2 \beta_i = (\delta^2\beta)_i = (\delta^2\beta)^{i-1} = \alpha^{i-1} , \quad \alpha = \delta^2\beta. \tag{9}$$

Let $\bar{x} = tu$, such that t is largest scalar for which $\bar{x}$ is feasible. Since

$$\beta u_{i-1} - u_i = \beta\alpha_{i-1} - \alpha_i = \beta\alpha_{i-1} - \alpha_{i-1}\beta \cdot \delta^2$$

$$= \beta\alpha_{i-1}(1-\delta^2) \le 0,$$

u will not hit the first (n-1) constraints, i.e. it will hit the n'th one.

Let t_i be such that

$$a_i t_i u = b_i \ , \quad i \le n.$$

Clearly $t_1 = 1$, for $i \ge 2$

$$t_i(\beta u_{i-1} + u_i) = b_i = \delta_i \ ,$$

$$t_i = \frac{\delta_i}{\alpha_{i-1}(\beta+\alpha)} \ ,$$

$$= \frac{\delta_i}{\beta\alpha_{i-1}(1+\delta^2)} = \frac{1}{(1+\delta^2)\beta} \frac{\delta^{i-1}}{(\delta^2\beta)^{i-2}} \ ,$$

$$t_i = \frac{1}{(1+\delta^2)\beta} \frac{\delta}{(\delta\beta)^{i-2}} \ .$$

Clearly,

$$t_1 .. t_i \ge t_{i+1} \ge \cdots \ge t_n = \frac{\delta}{\beta(1+\delta^2)} \frac{}{(\delta\beta)^{n-2}} \qquad (10)$$

Thus for $t = t_n$, u hits the n'th constraint. Hence $\bar{x} = tu$ is on the n'th hyperplane. Let us introduce slack variables so that

$$\begin{bmatrix} A & \begin{matrix} I_n & 0 \\ 0 & I_{n-1} \end{matrix} \end{bmatrix} \begin{bmatrix} \bar{x} \\ \bar{y} \\ \bar{z} \end{bmatrix} = b, \qquad (11)$$

$$\bar{y}_i = \delta_i - t(\beta\bar{x}_{i-1} + \bar{x}_i) , \quad \forall i \geq 2$$

$$= \delta_i - \frac{\delta}{\beta(1+\delta^2)} \frac{1}{(\delta\beta)^{n-2}} (\beta\alpha_{i-1} + \alpha_i)$$

$$= \delta_i - \frac{\delta}{(\beta+\alpha)} \frac{1}{(\delta\beta)^{n-2}} \alpha_{i-1}(\beta+\alpha)$$

$$= \delta_i - \frac{\delta}{(\delta\beta)^{n-2}} (\delta^2\beta)^{i-2} = \delta_i(1-(\delta\beta)^{-(n-i)} , \qquad (12)$$

$$\bar{y}_1 = 1 - t.$$

In particular $y_n = 0$. Let

$$\bar{z}_i = \bar{x}_i - \beta\bar{x}_{i-1} = t(\alpha_i - \beta\alpha_{i-1})$$

$$= t\alpha_{i-1}(\alpha-\beta) = t\alpha_{i-1}\beta(\delta^1-1)$$

$$= \frac{\delta}{\beta(1+\delta^2)} \frac{(\delta^2-1)\alpha_{i-1}}{(\delta\beta)^{n-2}} ,$$

$$\bar{z}_i = \delta \frac{(\delta^2-1)}{\delta^2+1} \frac{\delta_{i-1}\cdot\delta_{i-1}\cdot\beta_{i-1}}{(\delta\beta)^{n-2}} = \delta_i(\delta\beta)^{-(n-i)} \frac{\delta^2-1}{\delta^2+1} . \qquad (13)$$

Let D be dual of LP,

$$D \begin{cases} \min \sum_{i=1}^{n} u_i\delta_i \\ \text{s.t.} \quad u_1 + \beta u_2 + \beta\nu_2 \geq 1, \\ u_i + \beta u_{i+1} - \nu_i + \beta\nu_{i+1} \geq \beta^{i-1} , \quad 2 \leq i < n, \\ u_n - \nu_2 \geq \beta^{n-1} , \\ u \in R_+^n , \ \nu \in R_+^{n-1} . \end{cases}$$

Let $u^* = (0,0,\ldots,\beta^{n-1})$,

$\nu^* = (*,*,\ldots,*,0)$.

ν_i, $2 \le i \le n$ are determined from the u_i's using constraints of D. Let $x^* = (0,0,\ldots,0,\delta_n)$. Clearly,

$$cx^* = \Sigma u_i^* \delta_i = u_n^* \delta_n = \beta^{n-1} \delta_n = (\beta\delta)^{n-1} .$$

Thus at optimal

$$z_i^* = 0, \quad 2 \le i < 2 , \quad z_n^* = x_n^* = \delta_n,$$

$$y_n^* = 0, \quad y_i^* = \delta_i , \quad 1 \le i < n.$$

Let us write (11) in tableau form at Table 1. Let

$$\varepsilon_i = \delta_i (\delta\beta)^{-(n-i)} = \frac{\delta^{i-1}}{(\delta\beta)^{n-i}} = \frac{\delta^{2i-1}}{\delta^n \beta^n \beta^{-i}}$$

$$= \frac{1}{(\delta\beta)^n} \frac{\delta^{2i-1}}{\beta_i} = \frac{\theta(\delta\theta)_i}{(\delta\beta)^n} ,$$

$$\bar{z}_i = \varepsilon_i \phi , \quad \bar{z}_i \le \bar{z}_{i+1} \le,\ldots,\le \bar{z}_n ,$$

$$\bar{y}_i = \delta_i - \varepsilon_i ,$$

	x_1	x_2			x_n	y_1	y_2			y_n	z_2			z_n
1	1					1								
2	β	1					1							
3		β	1					1						
………														
			β	1					1					
n				β	1					1				
2	β	-1									1			
3		β	-1									1		
………														
			β	-1									1	
n				β	-1									1
X	t	$t\alpha$	$t\alpha^2$		$t\alpha^{n-1}$	$\bar{y}_1$	$\bar{y}_2$		$\bar{y}_{n-1}$	0	$\bar{z}_2$	$\bar{z}_3$		$\bar{z}_n$
C	1	β	β^2		β^{n-1}	0	0		0	0	0	0		0

Table 1

$$\phi = \frac{\delta^2-1}{\delta^2+1} .$$

Now c_n has largest (in absolute) value in C in Table 1. Since it is positive, we will increase x_n.

Let $X = (x,y,z) \in R^{3n-1}$; then

$$X^+(\tau) = \bar{X} + \tau\tilde{a}_n$$

$$= \bar{X} + \tau \begin{pmatrix} x_1 & & x_n & y_1 & y_n & z_2 & z_{n-1} \\ 0, & , & 1, & 0, & -1, & 0, & 1 \end{pmatrix} .$$

The only restriction is $y_n \geq 0$, but $\bar{y}_n = 0$, thus $\tau \quad 0$, so we will switch x_n with y_n in the basis, hence the new basis will be $\{x_n, y_1, .. y_{n-1}, z_2, .. z_n\}$.

The new tableau is (Tableau 2)

	x_1	x_2	x_3			x_n	y_1	y_2			y_n	z_2			z_n
1	1						1								
2	β	1						1							
3		β	1												
			β		1					1					
n					β	1					1				
2	β	-1										1			
3		β	-1										1		
			β		-1									1	
n					2β	0					1				1
X	t	$t\alpha$	$t\alpha^2$			$t\alpha^{n-1}$	y_1	y_2		y_{n-1}	0	z_2	z_3		z_n
C	1	β	β^2		ω_1	0	0	0		0	$-\beta^{n-1}$	0	0		0

Table 2

where $\omega_1 = \beta^{n-2}(1-\beta^2)$.

$$
x^2 = \begin{cases} \beta \bar{x}_{n-2} & \text{for} \quad x_{n-1} \\ \bar{x}_n + \beta \bar{z}_{n-1} & \qquad\; x_n \\ \bar{y}_{n-1} + \bar{z}_{n-1} & \qquad\; y_{n-1} \\ 0 & \qquad\; z_{n-1} \\ \bar{z}_n + 2\beta \bar{z}_{n-1} & \qquad\; z_n \\ x_i^1 & \qquad\; \text{otherwise.} \end{cases}
$$

But, since basic variable $z^2_{n-1} = 0$, we interchange x_{n-1} with z_{n-1}. Then $J = (x_{n-1}, x_n, y_1, \ldots, y_{n-1}, z_2, \ldots, z_{n-2}, z_n)$.

	x_1	x_2	⋮		x_n	y_1	y_2	⋮	y_n	z_2		⋮	z_n
1	1					1							
2	β	1					1						
3		β											
⋯	⋯	⋯	⋯	⋯	⋯	⋯	⋯	⋯	⋯	⋯	⋯	⋯	⋯
			1										
			2	0				1					
n					1				1				
2	β	-1								1			
		β	-1								1		
			$-\beta$	1								-1	
					0				1			2β	1
X	t		$t\alpha^{n-2}$	x_{n-1}	x_n	y_1	y_2	*	0	z_2	z_3	0	z_n
C	1		ω_2	0	0	0	0	0	$-\beta^{n-1}$	0	0	ω_1	0

Table 3

where $\omega_2 = \beta^{n-3}(1-\beta^2(\beta^2-1))$.

Since $C(x_{n-2}) < 0$, we will decrease x_{n-2}:

$$x^3 = x^2 + \tau(0 \quad -1, \beta, \beta^2, 0, 1, 2\beta, 0)$$

$$\tau = \min\{x_{n-2}, z_{n-2}, x_{n-1}/\beta\}.$$

$$z_{n-2} = x_{n-1} - \beta x_{n-2},$$

$$z_{n-2} \le x_{n-2},$$

$$x^2_{n-1} = x_{n-1} - z_{n-1} = \beta x_{n-2},$$

$$x^2_{n-1}/\beta = x_{n-2},$$

$$\tau = z_{n-2}.$$

$$x^3 = \begin{cases} \bar{x}_{n-2} - z_{n-2} & x_{n-2} \\ \bar{x}_{n-1} - \beta\bar{z}_{n-2} = 0 & x_{n-1} \\ \bar{x}_n + \beta^2\,\bar{z}_{n-2} & x_n \\ \bar{y}_{n-3} + \bar{z}_{n-3} & y_{n-3} \\ 0 & z_{n-2} \end{cases}$$

$$x^3(x_{n-1}) = 0$$
$$x^3(z_{n-2}) = 0$$

If we interchange z_{n-2} with x_{n-2}, we have a similar structure. Thus in (n-2) more steps, we will drive x_i to 0, for $i < n$. Thus we obtain the optimum solution.

4.15 LINEAR COMPLEMENTARITY PROBLEMS OF MURTY AND FATHI

The Linear Complementarity Problem (LCP), denoted by (M,q), with $q \in R^n$, $M \in R^{n\times n}$, is to find $x \in R^n$ satisfying

$$\text{LCP} \quad \begin{array}{l} Mx + q \geq 0 , \\ x \geq 0 \end{array} \qquad (1)$$

$$x^T(Mx + q) = 0. \qquad (2)$$

We will restrict ourselves to case where $M + M^T$ is positive semi-definite (PSD) matrix. Let

$$K = \{x\colon x \geq 0,\ Mx + q \geq 0\} ,$$

$$f(x) = x^T(Mx + q).$$

It is well-known [C22,C32], that when M is PSD, the Linear Complementarity Problem is solvable iff $K \neq \phi$. When CLP is solvable, there is a solution at an extreme point of K. Note that at any solution of LCP, $f(x) = 0$, and $f(x)$ is a convex function of x.

One can cast LCP as a convex programming problem:

$$\text{minimize } \{f(x) \,|\, x \in K\} .$$

Now consider the algorithm previously described. For $\tilde{x}_\nu \in K$, go to an extreme point $\underset{\sim}{x}_\nu$ of K, say, via projection method of quadratic programming, while minimizing $f(x)$. If $f(\underset{\sim}{x}) = 0$, $\underset{\sim}{x}$ is optimal; otherwise add the constraint

$$f(x) \leq f(\underset{\sim}{x}_\nu)$$

to (1) and solve the feasibility problem

$$P_\nu : \quad \begin{array}{l} Mx + q \geq 0 , \\ x \geq 0 , \\ f(x) \leq f(\underset{\sim}{x}_\nu), \end{array}$$

starting with $\tilde{x}_\nu$.

Katta G. Murty [M25], later Yahya Fathi [F5], exhibited a set of LCP with positive definite or semi-positive definite matrices and/or forms such that pivotal methods require 2^n or 2^n-1 pivots.

Let us define $M \in R^{n\times n}$ as

$$m_{i,i} = 1, \quad m_{i,j} = 0 \ , \ i < j \ , \quad m_{i,j} = 2 \ , \quad i > j \ ,$$

$$q^1 = -(1,\ldots,1),$$

$$q^2 = -(2^n, \ 2^n + 2^{n-1},\ldots,2^n + 2^{n-1} +\cdots+ 2).$$

Let

$$M_1 = M,$$

$$M_2 = MM^T \ ,$$

$$M_3 = M_2 + E,$$

$$M_4 = M_2 + F,$$

where $E_{i,j} = 4, \ \forall i,j \ , \ F_{i1} = 0, \ \ F_{ij} = -4, \ \ j > 1, \ \ \forall i.$
Let us define LCP_i^j, $i = 1,2,3,4, \ \ j = 1,2,$ as

$$\mathrm{LCP}_i^j: \begin{array}{l} M_i x + q^j \geq 0, \\ x \geq 0, \\ x^T(M_i x + q^j) = 0. \end{array}$$

Murty and Fathi each showed that LCP_i^1, $i = 1,\ldots,4$ will take 2^n steps with Lemke's complementary pivot method [L10]. LCP_i^2 takes 2^n-1 pivots with Murty's complementary pivot method [M25].

We will show that our algorithm solves all of above problems in two stages.

First let us give solutions. For $j = 1$, $x^* = (1,0,\dots,0)$. $w^* = M_i x + q^1 = (0,1,1,\dots,1)$. For $j = 2$, $x^* = (2^n,0,\dots,0)$. $w^* = (0,2^{n-1}, 2^{n-2}, \dots,2)$.

i) Starting with $x = 0$, the ellipsoidal algorithm will find a point in $K_i^j = \{x \in R^n : M_i x + q^j \geq 0\}$. If we denote by L_i^j the length of the binary coding of K_i^j as in [H 4 ,G13], it is obvious that $L_i^1 > n + n^2/2$, $L_i^2 > 3/2\, n^2$, $\forall i$. Thus the step length of any Hachiyan iteration will be greater than 1 for $j = 1$, and 2^n for $j = 2$, forcing the x^1 to be in K_i^j. Even if we do not consider the above fact, it is easy to see that in n iterations the algorithm will give a feasible point in respective K_i^j.

ii) Given any point in K_i^j, a vertex obtained by projection method via minimizing corresponding f(x) will give the unique solution.

Let us rewrite K_i^j as $K = \{x: Ax \geq b,\ x \geq 0\}$. It is clear that for $x \geq 0$,

$$a_1 x \geq b_1 \Rightarrow a_k x \geq 2a_1 x \geq 2b_1 > b_k \quad \text{for} \quad k > 1.$$

Thus in all cases $i = 1,2,\dots,4$, $j = 1,2$, only the first hyperplane can define a vertex of K. For $i = 1$, this forces $x_1 > 0$, hence $x_k = 0$, for $k > 1$. For other cases consider gradient of f(x):

$$g = 2Mx + q \geq Mx,$$

for $Mx + q \geq 0$. Since

$$a_1 x < a_2 x \leq a_3 x, \dots, \leq a_n x,$$

for any $x \geq 0$, $x \neq 0$ we have

$$g_1 < g_2 \leq g_3, \dots, \leq g_n, \tag{*}$$

thus forcing the last (n-1) components of x to zero. Although the formula for the gradient is not quite the same for $i = 4$, the relation (*) holds also in that case.

Thus we have shown that in all the cases $i = 1,\dots,4$, $j = 1,2$, the ellipsoidal algorithm will find a solution in K_i^j in at most n ellipsoidal iterations, and at most n projection iterations will give the required solution.

Recently Birge and Gana [B19] gave a class of LCP's with a P matrix so that van der Heyden's variable dimension and Dantzig-Cottle's principal pivoting method require 2^n-1 pivot to solve a problem of dimension n. Their example is obtained from Murty's example by transposing M, and reversing the order of elements in q. Since the basic geometry remains same, i.e., the underlying polyhedron has only one vertex, the above remarks equally apply to their problem, that is our algorithm solves this class almost instantly.

5 Convex programming via ellipsoidal algorithm

The ellipsoidal algorithm developed out of efforts for designing algorithms for convex minimization of nonsmooth functions with or without constraints (Shor [S18,S19]), and by Judin-Nemirovskii's work on informational complexity of general convex minimization problems [J7 ,J8].

In this chapter we study several convex programming problems in the light of works of Shor and Judin-Nemirovskii and give an estimation of complexity for them, namely: unconstrained minimization, general constrained minimization, linearly constrained minimization, convex quadratic, and the linear complementarity problem. We discuss its applicability to nonconvex programming.

5.1 GENERALITIES

ELLIPSOIDAL ALGORITHM vs. OTHER ALGORITHMS

A basic feature of subgradient algorithms in general, and ellipsoidal algorithm in particular is that they are not descent methods, in the sense that at each iteration the function value does not necessarily decrease. This is mainly because they decrease some other quantity, which is also strongly connected with the set of all optimizers of the given function. In subgradient and relaxation algorithms they minimize distance function d(x,P), where P is the solution set. In the ellipsoidal algorithm they reduce the volume of the region in which the optimizers are localized.

A second feature which is connected to the first is that we do not need any line search, since the step size is predetermined and independent of the direction at which the next point is located. Moreover, this direction is not a descent direction or a feasible direction.

The ellipsoidal algorithm is a finite, in particular polynomial, algorithm for some notion of approximate solution. As we mentioned earlier, there is no finite algorithm for solving general constrained convex minimization using a class of subgradient algorithms.

There is a vast literature on descent subgradient algorithms; in particular, Wolfe [W 9]; Lemarechal [L 8 ,L 9], Sandi [S 8], Gwinner [G 54] etc. (See [L 9] for a good reference list.)

The bounds on the number of iterations to achieve a certain approximation or to obtain an ε-optimal solution depend on f up to its Lipschitz constant, as an upper bound of subgradients, and is independent otherwise.

Preliminaries

We say $f: R^n \to R$ is convex if for $x_1, x_2 \in R^n$, $\lambda \in [0,1]$,

$$f(\lambda x_1 + (1-\lambda)x_2) \leq \lambda f(x_1) + (1-\lambda)f(x_2).$$

We say $u \in R^n$ is a subgradient of f at x, if

$$f(y)-f(x) \geq \langle u,y-x \rangle \ , \ \forall y \in R^n \ ,$$

or equivalently

$$f(x) \leq f(y) + \langle u,y-x \rangle \ .$$

The set of subgradients of f at x will be denoted by $\partial f(x)$, i.e.

$$\partial f(x) = \{u \in R^n : f(y)-f(x) \geq \langle u,y-x \rangle \ , \ \forall y \in R^n\}. \qquad (1)$$

If f is differentiable at x, then

$$\partial f(x) = \{\nabla f(x)\}.$$

If $f(x) = \max_{i \in I} f_i(x)$ and $I(x) = \{i \in I: f(x) = f_i(x)\}$, then

$$\partial f(x) = co\{\nabla f_i(x),\ \ i \in I(x)\}\ , \tag{2}$$

where co(A) denotes the convex hull of the set A.
In particular, when $f_i(x) = a_i x - b_i$,

$$\partial f(x) = \{u = \Sigma\{\lambda_i a_i: i \in I(\bar{x})\}: \lambda_i \geq 0,\ \Sigma\{\lambda_i: i \in I(\bar{x})\} = 1\}. \tag{3}$$

Such an f is, sometimes, called polyhedral.

The following is well-known for unconstrained minimization.

Theorem 1: If f is convex, x^* is a minimizer of f over R^n if and only if

$$\underline{0} \in \partial f(x^*). \qquad \square$$

For the constrained case

$$(P_2)\qquad \begin{array}{ll} & \min\ f(x) \\ \text{s.t.} & s(x) \leq 0, \end{array}$$

we consider the point to set map, $x \to M(x)$, where

$$M(x) = \begin{cases} \partial f(x) & \text{if}\quad s(x) < 0, \\ co(\partial f(x) \cup \partial s(x)) & \qquad s(x) = 0, \\ \partial s(x) & \qquad s(x) > 0. \end{cases} \tag{4}$$

Theorem 2: When f, s are convex, under some constraint qualifications, x^* solves P_2 if and only if

$$\underline{0} \in M(x^*).$$

Proof: See [S 3].

Definition: For $\varepsilon > 0$, $u \in R^n$ is said to be an ε-subgradient of f at x, i.e. $u \in \partial_\varepsilon f(x)$, if $f(y)-f(x) \geq \langle u,y-x\rangle - \varepsilon$, $\forall y \in R^n$. In other words,

$$\partial_\varepsilon f(x) = \{u \in R^n : f(y)-f(x) \geq \langle u,y-x\rangle - \varepsilon, \quad \forall y \in R^n\}. \tag{5}$$

Some references on ε-subgradiential calculus are Kutateladze [K 30-31], Hiriart-Urruty [H17].

Definition: A point $\bar{x}$ is called ε-optimal for f on R^n, if $f(\bar{x})-f(x^*) \leq \varepsilon$, where x^* is the optimal point for f, i.e.

$$f(x^*) = \inf_{x \in R^n} f(x) ,$$

in other words $f(y) \geq f(\bar{x})- \varepsilon$, $\forall y \in R^n$.

Theorem 3: $\bar{x}$ is ε-optimal for f $\iff \underline{0} \in \partial_\varepsilon f(\bar{x})$

Proof: See [K31, H17].

Although we will consider functions which are defined on R^n, we will study their restrictions on bounded, closed sets. In particular we will study them on ellipsoids.

Definition: $f: R^n \to R$ is called Lipschitz with constant $d > 0$ if

$$|f(x)-f(y)| \leq d \, ||x-y|| . \tag{6}$$

Given a compact S in R^n, a convex f is Lipschitz with constant d, where

$$d = \sup \{||u|| : u \in \partial f(x), \quad x \in S\}. \tag{7}$$

We will study the minimization problem over compact sets, mostly on ellipsoids. Either we will assume that the initial ellipsoid or euclidean ball is large enough so that it contains the optimal point when we are considering over all R^n, or we are interested in the restricted problem. The main reason for the latter is our inability to localize, a priori, the optimum solution even in general convex programming. Hachiyan [H 4] gave an example of this fact.

Let us recall some notation

$$E = (x,B) = \{x + By: ||y|| \le 1\}$$

will represent an ellipsoid with center x, and B is a non-singular transformation, $B: R^n \to R^n$. We will call $E_o = (x_o,B_o)$ the initial ellipsoid with radius $\le R$. g(x) will represent the cut or subgradient used at the point (x,B). Let d be such that

$$||g(x)|| \le d, \quad \forall x \in E_o. \tag{8}$$

Let

$$K(\varepsilon) = 4n^2 \log \frac{Rd}{\varepsilon},$$

and

$$M(\varepsilon) = 2n^2 \log \frac{Rd}{\varepsilon}.$$

We would like to recall two technical lemmas from Chapter 3. In the first, suppose $H \subset G$, H, G are convex, compact subsets of R^n. If $|H| / |G| \le \omega^n$, then for any $\bar{x} \in H$, $\exists$ a direction e, $||e|| = 1$, such that the ratio of line segments originating from $\bar{x}$ in the direction e will not exceed ω. In the second one, thickness of a compact set $\rho(H) \le |H|^{1/n}$, where $|H|$ is determined so that volume of the unit ball in R^n is 1.

The ellipsoidal algorithm, as mentioned earlier, is a particular element of the subgradient family using a space dilatation operator. Let us recall that for $||s|| = 1$,

$$R_\alpha(s): x \to x + (\alpha-1) \langle x,s\rangle s, \tag{9}$$

$$B^+ = B R_\beta(s), \quad \beta = \alpha^{-1}, \tag{10}$$

$$x^+ = x - \gamma BB^t g/||B^t g|| \ . \tag{11}$$

By letting $H = BB^t$, we can rewrite the above update formulas as

$$H^+ = H - (1-\beta^2) \frac{(Hg)(Hg)^t}{g^t Hg} \ . \tag{12}$$

By defining $||x||_H^2 = ||x^t Hx||$, we can write the update formula (11) as

$$x^+ = x - \frac{\gamma (Hg)(Hg)^t}{||g||_H} \ . \tag{11'}$$

In this sense this whole class can be seen as a variable metric method with a rank-one update.

Let us recall some properties of $R_\alpha(s)$.

<u>Lemma 4</u>: $\alpha > 0$, $\beta > 0$, and $||s|| = 1$, we have:

(i) $R_{\alpha\beta}(s) = R_\alpha(s) R_\beta(s)$.

(ii) $||R_\alpha(s)x||^2 = ||x||^2 + (\alpha^2-1) <x,y>^2$.

(iii) Eigenvalues of $R_\alpha(s)$, are α , and 1 with multiplicity $(n-1)$.

(iv) $\det R_\alpha(s) = \alpha$.

(v) $\min(\alpha,1)||x|| \leq ||R_\alpha(s)x|| \leq \max(\alpha,1)||x||$. □

The following is a simple consequence of the definition:

<u>Lemma 5</u>: $H_k = B_k B_k^t$ is positive definite $\forall k$. □

Since $(Hg)(Hg)^t$ is rank-one, symmetric, and, hence positive semidefinite, its trace is positive. Thus trace, $\mathrm{tr}(1-\beta^2)(Hg)(Hg)^t/g^t Hg = \delta$ is positive. Since

$$\mathrm{tr}\ H^+ = \mathrm{tr}\ H - \delta \ ,$$

we have at each iteration

$$\text{tr } H^+ < \text{tr } H.$$

Consequently,

$$0 < \text{tr } H_{k+1} < \text{tr } H_k < \cdots < \text{tr } H_o = \text{tr } I = n.$$

Thus, at each iteration, trace of the update matrix is positive and less than n. Thus we have

<u>Lemma 6</u>: [Y 1] $\text{tr}(1-\beta^2) \dfrac{(Hg)(Hg)^t}{g^t Hg} < n$. □

5.2 UNCONSTRAINED MINIMIZATION OF A CONVEX FUNCTION OVER AN ELLIPSOID

Consider the problem

$$CP_1: \quad \min_{x \in E_o} f(x)$$

If E_o contains any minimizer point x^*, then it is equivalent to $\min\{f(x): x \in R^n\}$. Otherwise we are interested in the restricted problem i.e., its restriction to E_o.

The following is proved in Chapter 3, and taken from Shor [S19].

<u>Theorem 1</u>: The ellipsoidal algorithm gives a point $\bar{x}$ such that

$$|\langle g(\bar{x}), \bar{x} - x^* \rangle| < \varepsilon$$

in $K(\varepsilon)$ iterations, where x^* is a solution of the problem.

<u>Corollary 2</u>: The ellipsoidal algorithm finds an ε-optimal solution for CP_1 in at most $K(\varepsilon)$ iterations.

<u>Proof</u>: Follows from Theorem 1. Then $x = \text{argmin } \{f(x_\ell): \ell \le K(\varepsilon)\}$ is an ε-optimal solution.

Recently J.L. Goffin [G35] has obtained a slightly better form of the above corollary and similar corollary of the next section.

We would like to point out that the rate of convergence is a worst case bound, but it is quite possible to obtain at least a similar rate.

First we will consider a very simple example: a linear maximization over the unit ball.

$$f(x) = e_1^t x \text{ , over } E_o = \{x: ||x|| \le 1\}$$

starting with $x = 0$. Following Shor's development,

$$g(x) = e_1, \quad \forall x \in E_o \text{ ,} \quad s_1 = e_1 \text{ ,} \quad B_1 = R_\beta(s_1) \text{ ,}$$

$$s_k = e_1, \quad B_k = R_\beta(s_1),\ldots,R_\beta(s_k) = R_\beta k(e_1) \text{ ,}$$

$$h_k = \frac{\sigma^k}{n+1} \text{ ,} \quad \tau_k = h_k \beta^k \text{ ,}$$

since

$$B_k^t \, g(x)/||B_k^t \, g(x)|| = e_1,$$

and

$$B_k \, s_k = B_k \, e_1 = \beta^k \, e_1 \text{ .}$$

Thus

$$\tau_k = h_k \beta^k = (\frac{1}{n+1} \, (\frac{n^2}{n^2 - 1} \cdot \frac{n-1}{n+1})^{k/2} = \frac{1}{n+1} \, (\frac{n}{n+1})^k = \frac{1}{n+1} \, \rho^k \text{ ,}$$

where $\rho = \frac{n}{n+1}$,

$$x_{k+1} = x_k + \tau_k \, e_1 \text{ ,}$$

$$x_{k+1} - x_k = \frac{1}{n+1} \, \rho^k \, e_1,$$

$$e_1^t x_{k+1} = \frac{1}{n+1} \sum_{\ell=0}^{k} \rho^{\ell} = \frac{1}{n+1} \frac{1-\rho^{k+1}}{1-\rho} = 1 - \left(\frac{n}{n+1}\right)^{k+1} = 1 - \rho^{k+1} ;$$

thus, $||x_k - x^*|| = \rho^{k+1}$.

In order to have

$$|f(x_k) - f(x^*)| = ||x_k - x^*|| = \rho^{k+1} \leq \varepsilon ,$$

we need $k+1 \approx \dfrac{|\log\varepsilon|}{|\log\rho|} = \dfrac{|\log\varepsilon|}{|\log n/n+1|}$.

Let us note that we have

$$\lim_{k \to \infty} x_k = x^* .$$

For $n = 1$, the ellipsoidal algorithm reduces to Levin's algorithm, which in turn, becomes the ordinary binary search. In this case, we have

$$x_k = \frac{1}{2} [1 + \frac{1}{2} + \cdots + (\frac{1}{2})^k] = 1 - (\frac{1}{2})^{k+1} .$$

Now, let $h: R \to R$ be convex monotonic increasing function, and $f(x) = h(e_1^t x)$. Then the above analysis yields the condition

$$|f(x_k) - f(x^*)| \leq d R \rho^{k+1} .$$

In order to guarantee that this quantity will be $\leq \varepsilon$, we need

$$k+1 \approx \frac{|\log\varepsilon/Rd|}{|\log\rho|} .$$

This example shows that the worst case bound can easily be attained. It also demonstrates the importance of Theorem 3.6.2.

5.3 CONSTRAINED MINIMIZATION OF A CONVEX FUNCTION OVER AN ELLIPSOID

Let the convex programming problem be

$$CP_2' \qquad \min_{x \in E_o} f(x)$$

$$\text{s.t.} \quad f_i(x) \le 0, \quad i = 1,\ldots,m.$$

CP_2' is equivalent to

$$CP_2 \qquad \min_{x \in E_o} f(x)$$

$$\text{s.t.} \quad s(x) \le 0,$$

where $s(x) = \max_i f_i(x)$. Let

$$G_\varepsilon(x) = \begin{cases} \partial s(x) & , \quad s(x) > \varepsilon \\ \partial f(x) & , \quad s(x) \le \varepsilon. \end{cases}$$

Let $S(\varepsilon) = \{x: s(x) \le \varepsilon\}$. We will assume that $S(0) = \phi$ implies $S(\varepsilon) = \phi$. Equivalently $s(x^*) > 0$ implies $s(x^*) \ge 2\varepsilon$, where $s(x^*) = \min_{x \in E_o} s(x)$.

Definition: When CP_2 is feasible, $\bar{x}$ is an ε-optimal if $\max \{|f(\bar{x})-f(x^*)| \ , \ s(\bar{x})\} \le \varepsilon$. When CP_2 is infeasible, $\bar{x}$ is ε-optimal, if $s(\bar{x})-s(x^*) < \varepsilon$.

This definition is very similar to the definition of $*$-defect of Judin-Nemirovskii [J 7]. It is also similar to the definition of weak optimizer of Grötschel-Lovász-Schrijver [G 48].

Let $M(\varepsilon) = 2n^2 \log \frac{Rd}{\varepsilon}$. The following is well-known.

Theorem 1: The ellipsoidal algorithm gives an ε-optimal solution to CP_2 in at most $M(\varepsilon)$ iterations with $g(x) \in G_\varepsilon(x)$.

Corollary 2: If $\{\ell: s(x_\ell) \le \varepsilon\} \ne \phi$, let $\bar{x} = \text{argmin}\{f(x_\ell)|s(x_\ell) \le \varepsilon,\ \ell \le M(\varepsilon)\}$. Otherwise let $\bar{x} = \text{argmin}\{s(x_\ell):\ \ell \le M(\varepsilon)\}$. Then $\bar{x}$ is an ε-optimal solution for CP_2.

It is obvious that the result of this section contains that of the last section. In fact one obtains the following

Corollary 3: The ellipsoidal algorithm finds an ε-optimal solution of CP_1 in $M(\varepsilon)$ steps by choosing $s(x) \equiv 0$.

The perturbation of constraints is not new in mathematical programming. Rockafellar, used it in a theoretical level to obtain some deep theorems in convex programming. A.R. Conn used an ε-active strategy to avoid zig-zagging and obtained better performance (see [C28], for example). The perturbation introduced by Judin-Nemirovskii is of a different character, which stems from the geometry of convex sets and convex functions.

5.4 MINIMIZATION OF A CONVEX FUNCTION SUBJECT TO LINEAR INEQUALITIES

Consider the following problem

$$CP_3 \qquad \begin{array}{l} \min f(x) \\ \text{s.t.} \quad Ax \le b \end{array}$$

where $A \in Z^{m\times n}$, $b \in Z^m$ $f: R^n \to R$.

Let L be the length of binary encoding of data for A, defined earlier in Chapter 3.

Let us recall some of the lemmas from Chapter 3 and Chapter 4.

Lemma 1: If $K = \{x: Ax \le b\}$ is non-empty, then every vertex of K is a rational vector of which every component has denominator and nominator bounded by $2^L/mn$. □

Lemma 2: Let $s(x) = \max_i \; a_i x - b_i$. If $K = \phi$, then

$$s(x) \geq 2 \cdot 2^{-L}, \; \forall x \in R^n . \qquad \square$$

Lemma 3: If $s(\bar{x}) < 2^{-L}$ for some $\bar{x}, \exists \tilde{x}$ s.t. $s(\tilde{x}) = 0$, i.e. $\tilde{x}$ is feasible. Moreover $\tilde{x}$ can be obtained from $\bar{x}$ by projection method or pivoting in polynomial time, via VERTEX. $\square$

Thus one can take as an initial ellipsoid

$$E_o = (0, 2^L I).$$

Let

$$\varepsilon_1 = 2^{-L}, \; L_1 = \max \{L, \lceil \log d \rceil\}$$

where d is such that

$$\forall g \in \partial s(x), \; ||g|| \leq d, \quad x \in K_\varepsilon$$

and

$$\forall g \in \partial f(x), \; ||g|| \leq d, \quad \forall x \in K_{\varepsilon_1}$$

where

$$K_{\varepsilon_1} = \{x: Ax \leq b + e \cdot \varepsilon_1\}, \quad e = (1,1,\ldots,1) \in R^m .$$

For

$$\varepsilon \leq \varepsilon_1 ,$$

let

$$M(\varepsilon) = 2n^2(|\log\varepsilon| + L_1 + L).$$

Thus $\bar{x}$ of Corollary 3.2 is ε-optimal and can be obtained in at most $M(\varepsilon)$ iterations by the ellipsoidal algorithm.

Let B be any basis,

$$||B^{-1}(b + \varepsilon \cdot \bar{1}) - B^{-1} b|| \leq ||B^{-1}\varepsilon \cdot \bar{1}|| \leq \frac{2^L}{m \cdot n}\sqrt{n}\varepsilon = \frac{2^L}{m\sqrt{n}} \varepsilon = \delta.$$

$$f(B^{-1}b+\varepsilon\cdot\bar{1})-f(B^{-1}b) = \langle f'(B^{-1}(b + t\varepsilon\cdot\bar{1})), B^{-1}\cdot\varepsilon\cdot\bar{1}\rangle$$

for some $t \in (0,1)$ by the mean value theorem.
Thus

$$|f(B^{-1}b + \varepsilon\cdot\bar{1})-f(B^{-1}b)| \leq d\,\delta = \theta.$$

Given $\bar{x}$ such that $|f(\bar{x})-f(x^*)| \leq \varepsilon$, $s(\bar{x}) \leq \varepsilon$. Let $\tilde{x}$ be obtained from $\bar{x}$ by minimizing $f(x)$ over $Ax \leq b + \varepsilon\bar{1}$, say, by projection. In at most n iterations, if $\tilde{x}$ is a vertex of K_ε, then $\underset{\sim}{x}$ is the corresponding vertex of K, and by the above computation, we have

$$||f(\tilde{x})-f(\underset{\sim}{x})|| \leq \theta ,$$

which implies that

$$f(\underset{\sim}{x})-f(x^*) \leq \theta + \varepsilon.$$

If $\tilde{x}$ is not a vertex, let $\underline{x}$ be its projection on K. Then $\underline{x}$ is feasible, and

$$f(\underline{x})-f(x^*) \leq \theta + \varepsilon.$$

When the problem has linear equalities as constraints, one can eliminate some of the variables and work in the null space of the system of equalitions. For a numerically stable way of achieving this, see for example [G24, G25, G40].

5.5 CONVEX QUADRATIC PROGRAMMING

Consider

$$\text{QP} \quad \min f(x) = \frac{1}{2} x^t Cx + dx \tag{1}$$

$$\text{s.t.} \quad Ax \leq b.$$

Let us define

$$L_1 = [\sum_{i,j-1}^{n} \log_2(|c_{ij}| + 1) + \sum_{j=1}^{n} \log_2(|d_j| + 1)],$$

$$L_2 = [\sum_{i,j=1}^{mn} \log(|a_{ij}| + 1) + \sum_{j} \log(|b_j| + 1) + \log_2 mn + 1],$$

$$L = L_1 + L_2.$$

Let us write the Kuhn-Tucker condition for QP

$$Cx + d = A^T\lambda,$$

$$\lambda \geq 0, \tag{2}$$

$$Ax \leq b,$$

$$\lambda^t (Ax - b) = 0. \tag{3}$$

Considering the above equalities/inequalities, we know that for any (x,λ) satisfying the above requirements there is one which satisfies vertices of the corresponding polyhedron (2). Thus it suffices to check the defining equations/inequalities for the reduced system (2). Since the corresponding length of the binary encoding $L' \leq 2L$, any solution of (x,λ) will have the form t/s, where $|t|, |s| \leq 2^{2L}$. Since the objective function is quadratic, the smallest rational number will be $(1/2^{2L})^2 = 1/2^{4L}$. Hence

$$||\frac{1}{2} x^T Cx + dx|| \leq ||x||^2||C|| + ||d|| \, ||x|| \leq$$

$$\leq 2^L \cdot (2^{2L}) = 2^{5L} .$$

Thus $|f(x^*)| \leq 2^{5L}$. Combining with the previous observation $|f(x^*)| = t/s$, when t and s are relatively prime, $|t| \leq 2^{5L}$, $|s| \leq 2^{4L}$.

But, as we previously mentioned, by a well-known theorem [H16, R3] it suffices to approximate $f(x^*)$ by a real (or rational) number r which satisfies

$$|f(x^*)-r| < \frac{1}{2(2^{4L})^2} = 2^{-8L-2} .$$

Kozlov, Tarasov and Hachiyan [K22] reduce the solution of QP to solution of a sequence of problem of type

$$P_k: \quad Ax \leq b,$$
$$f(x) \leq t_k/s_k ,$$

where $t_k, s_k \in Z$, $k = 1,\ldots,13L + 2$.

They solve each P_k by the ellipsoidal algorithm for decision problem in $0(n^2L)$ iterations. They first show that $f(x^*)$, the optimum value of the objective function can be represented as t/s where s and t are relatively prime, with $|t| \leq 2^{5L}$, and $|s| \leq 2^{4L}$. Then in order to find $f(x^*)$ exactly, it suffices to find $\bar{x}$ such that

$$f(\bar{x})-f(x^*) \leq 2^{-8L-2} .$$

Then they suggest to compute $f(x^*)$ by expanding $f(\bar{x})$ in continued fractions [H16, R3]. Clearly this can be done in polynomial time. In general, if $f(\bar{x})-f(x^*) < 1/D$, $f(x^*) \leq f(\bar{x})$, $f(x^*)$ can be calculated as $1/D \lfloor f(\bar{x})D \rfloor$ provided that we have the ability to do such calculations. Then an exact solution of x^* can be obtained by solving:

$$Ax \le b$$

$$f(x) \le f(x^*),$$

by means of a series of decision problems (see also [P 6]).

Instead of solving a sequence of P_k one can apply a direct ellipsoidal algorithm with $g(x) \in G_\varepsilon(x)$, where

$$G_\varepsilon(x) = \begin{cases} \partial f(x) & \text{if} \quad s(x) \le \varepsilon \\ \partial s(x) & \phantom{\text{if}} \quad s(x) > \varepsilon . \end{cases}$$

Let $\varepsilon_1 = 2^{-L_2}$, $\epsilon = 2^{-8L-2}$. One can easily verify that

<u>Lemma 1</u>: $f'(x) = Cx + d$, $||f'(x)|| \le h = 2^L$. $\forall x \in R^n$, such that

$$||x|| \le 2^{L_2} .$$ □

Thus from the results on general convex programming, we have

<u>Theorem 2</u>: The ellipsoidal algorithm gives a point $\bar{x}$ such that

$$|f(\bar{x})-f(x^*)| < \varepsilon \quad , \quad s(\bar{x}) < \varepsilon ,$$

in

$$K(\varepsilon) = 2n^2(|\log\varepsilon| + \log h + \log R) \le 20n^2 \cdot L$$

with the condition that

$$g(x) \in G_{\varepsilon_1}(x).$$

5.6 LINEAR COMPLEMENTARITY PROBLEM

Consider the problem of finding x such that

$$x \geq 0$$

$$Mx + q \geq 0$$

$$\text{LCP} \quad x^t(Mx + q) = 0$$

where $M \in R^{n \times n}$.

Let $f(x) = x^t(Mx + q)$. And let $S = \{x \in R^n : x \geq 0, Mx + q \geq 0\}$. Let L be length of binary encoding of S:

$$L = [\Sigma \log(|m_{ij}| + 1) + \Sigma \log(|q_j| + 1) + \log n + 1].$$

Lemma 1: If x is a vertex of S, then

$$x_i \neq 0 \Rightarrow x_i \geq 2^{-L} n,$$

$$(Mx + q)_i \neq 0 \Rightarrow (Mx + q)_i \geq 2^{-L} n. \qquad \square$$

Lemma 2: If LCP has a solution, then it has one at a vertex of S. Moreover at every solution of LCP, $f(x) = 0$. □

Proof: See [M 7].

Lemma 3: When M is positive semi-definite (PSD), then LCP has a solution $\Leftrightarrow S \neq \phi$.

Proof: See [C 32].

It is well-known that the quadratic programming problem and the linar programming problem can be written as LCP. But here we will treat LCP as a convex quadratic programming problem. We will assume that f is convex, or equivalently $M + M^t$ is PSD.

By following the recipe given in Kozlov, Tarasov and Hachiyan one can find on exact solution of

$$x \geq 0,$$

$$Mx + q \geq 0,$$

$$x^t(Mx + q) \leq 0,$$

in (2n) applications of the procedure described in Section 4.1.

Let

$$L_1 = \max \{L, \lceil \log_2 d \rceil \},$$

where $||2Mx + q|| \leq d, \quad \forall x \in E_o = \{x: ||x|| \leq 2^L\}$.

Lemma 4: If vertex x of S is not a solution to LCP, then $f(x) \geq n^2 \, 2^{-2L}$. □ Let $\varepsilon_1 = 2^{-L}$, $\varepsilon_2 = \frac{1}{2} \, 2^{-2L} \, n^2$. Let

$$S_\varepsilon = \{x: -x \leq \varepsilon, \; -Mx \leq -q + \bar{1} \cdot \varepsilon\}.$$

Let

$$\varepsilon = \varepsilon_2, \quad M(\varepsilon) = 2n^2(|\log\varepsilon| + L_1 + L).$$

Theorem 5: Ellipsoidal algorithm will give an $\bar{x}$ s.t. $f(\bar{x}) < \varepsilon$, $s(\bar{x}) < \varepsilon$ in atmost $M(\varepsilon)$ iterations with $G_\varepsilon(x)$ as set of sub-gradients. □

Let $P = \{x: Ax \leq b\}$, $P_\varepsilon = \{x: Ax \leq b + \bar{1} \cdot \varepsilon\}$. Then there is a well-defined correspondence between the vertices of P and P_ε. Let B be a basis of A, corresponding to a vertex of P i.e.

$$\underset{\approx}{x} = B^{-1} b .$$

Then $x = B^{-1}(b + \bar{1} \cdot \varepsilon)$, will give a vertex of P_ε, and conversely. Note that non-negativity constraints (if any) are dealt with implicitly.

It is quite easy to see that $\{x: f(x) < \varepsilon, \; s(x) < \varepsilon\}$ contains only those vertices of S which are the solutions of LCP.

Since

$$f(x) = x^t Mx + q^t x ,$$

$$f(x + \delta) = f(x) + f'(x)^t \delta + \delta^t M\delta .$$

In particular, for $\delta = \tilde{x} - \underset{\approx}{x}$

$$f(\tilde{x})-f(\underset{\approx}{x}) = \delta((M + M^t)\underset{\approx}{x} + q) + \frac{1}{2}\delta(M + M^t)\delta .$$

So

$$|f(\tilde{x})-f(\underset{\approx}{x})| \leq ||g(\underset{\approx}{x})|| \, ||\delta|| + \frac{1}{2} ||M + M^t|| \, ||\delta||^2$$

$$\leq d||\delta|| + d||\delta||^2 / 2 .$$

Since $\delta = B^{-1}\varepsilon \, \bar{1}$, $||\delta|| \leq 2^L \cdot \varepsilon < 1$, for ε sufficiently small we have

$$||\delta||^2 < ||\delta|| .$$

Thus

$$|f(\tilde{x})-f(\underset{\approx}{x})| \leq 2d||\delta|| \leq 2 \cdot 2^{L_1} \cdot 2^L \cdot \varepsilon$$

for $\varepsilon = \frac{1}{2} 2^{-(3L+L_1)}$, we have

$$|f(\tilde{x})-f(\underset{\approx}{x})| < 2^{-2L} .$$

On the other hand

$$|f(\bar{x})-f(x^*)| < \varepsilon < 2^{-2L} ,$$

$$f(\underset{\approx}{x}) \leq f(\tilde{x}) + 2^{-2L} \leq f(\bar{x}) + 2^{-2L} \leq f(x^*) + 2^{-2L} + 2^{-2L} .$$

$$\therefore \quad |f(\underset{\approx}{x})-f(x^*)| < 2 \cdot 2^{-2L}$$

$$\Longrightarrow f(\underset{\approx}{x}) = f(x^*).$$

The linear complementarity problem has been studied via ellipsoidal algorithm. Jones and Marvil [J 6] studied the positive definite case. Chung and Murty [C22], and Adler-Maclean and Provan [A2] studied the positive semidefinite case. The methods in [A2] and [C22] are based on Gács-Lovász approach, and both assume that M is positive semidefinite. The approach in [J6] is slightly different and uses the properties of positive definite matrix M. As far as we know, the above methods cannot be modified to cover the case when $M + M^t$ is positive semidefinite, which we proved above.

5.7 QUASI-CONVEX PROGRAMMING

A function $f: R^n \to R$ is quasi-convex set S if the sets $S_\alpha = \{x: f(x) \le \alpha\}$ are convex $\forall \alpha \in R$. Equivalently, f quasi-convex on S if $x \in S$, $f(x) \le f(\bar{x})$, $\lambda \in [0,1]$ implies

$$f(\lambda x + (1 - \lambda)\bar{x}) \le f(\bar{x}). \tag{1}$$

Furthermore, if f is differentiable, then quasi-convexity implies that

$$f(x) \le f(\bar{x}) \Longrightarrow \nabla f(\bar{x})(x - \bar{x}) \le 0. \tag{2}$$

When f is differentiable, the inequality

$$\nabla f(x)(x - x^*) \ge 0 \tag{3}$$

is a valid cut in view of (2) with $f(x^*) \le f(x)$.

Thus the ellipsoidal algorithm can be applied to quasi-convex programming problems.

For a quasi-convex function a local minimizer is not necessarily a global minimizer. However, when f is pseudo-convex a local minimizer is also a global minimizer. Let us recall that a function f is pseudo-convex if it is differentiable and

$$\nabla f(\bar{x})(x - \bar{x}) \geq 0 \Rightarrow f(x) \geq f(\bar{x}). \quad (4)$$

A pseudo-convex function is necessarily quasi-convex.

After the initial localization of the required solution, we need two essential feature from the function which we want to minimize:

(i) any local minimum or stationary point must be a global minimum, and

(ii) at any point, we can either produce a cut separating the set of solutions on the one side of cut, or prove that the current point is a stationary point.

In this respect those classes of functions for which any stationary point is a global minimum become important. These classes functions are characterized by Zang, Choo and Avriel in a series of papers ([Z 3 - 5]). We will summarize their results. See also Horst [H21] .

For f: $R^n \to R$, let $R_f = \{f(x)\ x \in R^n\} = f(R^n)$ be the range of f. For $\alpha \in R_f$, let $L_\alpha(f) = \{x \in R^n : f(x) \leq \alpha\}$. The functions in question are characterized by the continuity properties of the point to set mapping $\alpha \to L_\alpha(f)$, for $\alpha \in R_f$.

<u>Definition</u>: The point to-set mapping $L_\alpha(f)$ is said to be lower semicontinuous (ℓsc) at $\alpha \in R_f$, if $x \in L_f(\alpha)$, $\alpha_i \in R_f$, $(\alpha_i) \to \alpha$, implies $\exists$ K and $(x^i) \to R^n$ such that for $i \geq K$, $x^i \in L_f(\alpha^i)$, and $x^i \to x$.

<u>Theorem 1</u>: [Zang and Avriel] Let f: $C \subseteq R^n \to R$. Every local minimum of f on C is a global minimum on C iff $L_f(\alpha)$ is lower semi-continuous.

<u>Definition</u>: f: $C \subseteq R^n \to R$ is called <u>strictly quasi-convex</u> on C if $\tilde{x} \in C$, $\bar{x} \in C$, $f(x) < f(\bar{x})$, $\lambda \in (0,1)$ implies that

$$f(\lambda \tilde{x} + (1 - \lambda)\bar{x}) < f(\bar{x}).$$

Theorem 2: [Zang and Avriel] Let f be strictly quasi-convex function on a convex set $C \subseteq R^n$. Then $L_f(\alpha)$ is ℓsc on R_f. □

For functions whose stationary points are global minimum we need a stronger

Definition: The mapping $L_f(\alpha)$ as called strictly lower semi-continuous (sℓsc) at $\alpha \in R_f$ if $x \in L_f(\alpha)$, $\alpha_i \in R_f$, $\alpha_i \to \alpha$ implies that $\exists$ K, a sequence (x^i), and $\beta(x) > 0$ such that $x_i \in L_f(\alpha_i - \beta(x)||x - x^i||)$, for $i \geq K$, and $x^i \to x$.

Zang, Choo and Avriel developed a theory to characterize the functions whose stationary points are global minima using the concept of cone of tangents developed by Hestenes. For a differentiable function it reduces to

Theorem 3: Let f be a real-valued differentiable function on an open set containing $C \subseteq R^n$. Every stationary point of f is a global minimum of f on C iff $L_f(\alpha)$ is sℓsc on R_f. □

Consider the problem

$$\text{PP} \qquad \begin{array}{l} \min f(x) \\ \text{s.t. } f_i(x) \leq 0, \quad i = 1,\dots,m \end{array}$$

and let

$$G_\varepsilon(x) = \begin{cases} \nabla f(x) & \text{if} \quad \max_i \{f_i(x)\} \leq \varepsilon \\ \nabla f_i(x) & \text{if} \quad f_i(x) > \varepsilon . \end{cases}$$

Suppose we applied the ellipsoidal algorithm to PP with $g(x) \in G_\varepsilon(x)$. Then in at most $M(\varepsilon)$ iterations, we reach a point where conditions for Theorem 3.6.5 are satisfied except the convexity of functions. We will follow that proof in principle. If $x^* \in H$, (last ellipsoid except the trivial parts), then by Lemma 3.6.1 there exists y, $z \in E_o$ such that

$$\frac{||x^* - y||}{||x^* - z||} \le \omega \tag{5}$$

i.e. $||x^* - y|| \le \omega ||x^* - z|| \le 2\omega \cdot R$.

On the other hand, by the mean value theorem

$$f(x^*) - f(y) = \langle \nabla f(x^* + t(y - x^*)),\ x^* - y \rangle$$

for some $t \in (0,1)$. Thus

$$|f(x^*) - f(y)| \le 2\ \omega \cdot R \cdot d\ . \tag{6}$$

But by assumption $\omega \cdot R \cdot d = \varepsilon$. Thus we have

$$f(x^*) \ge a - 2\omega R d = a - 2\varepsilon\ , \tag{7}$$

where $a = \min\ \{f(x_\ell) : \ell \le x,\ \max_i f_i(x_\ell) \le \varepsilon\}$.
Hence we have proved:

<u>Theorem 4</u>: When the ellipsoidal algorithm is applied to PP with $g(x) \in G_\varepsilon(x)$, in $M(\varepsilon)$ iterations we find $\bar{x}$ such that $s(\bar{x}) \le \varepsilon$, and $f(x^*) \ge f(\bar{x}) - 2\varepsilon$, where x^* is global minimizer, provided f, f_i's are pseudo-convex functions. When f is quasi-convex, then x^* is a local minimizer. When f_i's are quasi-convex and PP is feasible, it is necessary that $\{i: f_i(x) > \varepsilon,\ \nabla f_i(x) \ne 0\} \ne \phi$, when $s(x) > \varepsilon$ at every point that algorithm generates; otherwise the algorithm is not valid.

<u>Corollary 5</u>: For unconstrained PP, i.e. $s(x) \equiv 0$, the ellipsoidal algorithm finds an ε-optimal solution for a pseudo-convex function, and a local ε-optimal solution for a quasi-convex function in $M(\varepsilon)$ iterations.

5.8 NUMERICAL RESULTS AND FURTHER COMMENTS

Only a limited amount of experiments have been reported in the literature. In the pre-Hachiyan period, Shor's space dilatation with difference of successive gradient [S27] was reported to have better performance than Wolfe's [W 6], and Lemarechal [L8] methods, in Shor [S15]. This view is also shared by Lemarechal (see [L 9]).

Goffin [G30] reported some test results along with an incomplete, inconclusive theory. He reports the practical convergence rate of the ellipsoidal algorithm reduced to $\approx 1-\frac{.7}{n}$ much below the theoretical level $q^{1/n} \approx 1-\frac{1}{2n^2}$.

Ecker and Kupferschmid [E 2] performed a rather extensive comparison with a commercial code: NAG. Suprisingly, they report that the ellipsoidal algorithm is competitive with the commercial NAG code. The problems they considered contain 20-30 variables. They also reported that they had no difficulty in implementing Gács-Lovász version from the stability point of view.

Barton [B10] reported that he solved a traffic assignment problem with the ellipsoidal algorithm. The algorithm outperformed the competitive Wolfe [W6], Lemarechal [L 8], and Shor [S 15] algorithms.

One would expect that one could apply the ellipsoidal algorithm in a more general context. We suspect that it can be used with Clarke's [C 23, 24] generalized gradient, or Shor's almost gradient [S16] and many similar ones (see also Ioffe [I 2]).

Since the ellipsoidal algorithm is a certain rank-one variable metric algorithm, one would wonder what the limiting behaviour of B_k, or H_k will be. some partial answers are known: Goffin [G 33] for the solution of linear equality system with cyclic implementation, showed that there are n limiting shapes of theellipsoids. This seems an area for further research, along with Shor's algorithm [S 15]. Some efforts has been made (see [Y2 , G30]). For the relation between the ellipsoidal algorithm and variable metric algorithms, and for the interpretation of the final form of the ellipsoid as a certain "Hessian", see Goffin [G 37].

Recently, Bakan et al [B6-7] applied ellipsoidal algorithm to a system of strict inequalities with uncountable constraints. With the assumption that an a priori lower bound on the volume of the solution set is known, they showed that in polynomial time one can find an ε-feasible solution. In fact one can do more, find a feasible solution. It seems that strict linear inequalities are natural in pattern recognition; see B6-7 and reference given there.

For some recent numerical experiments in Soviet Union, see [S2, S34, M18].

6 Solving convex–concave games via ellipsoidal algorithm

In this chapter we study the optimization of convex-concave functions, and in particular convex-concave games over simplices or arbitrary polyhedral sets.

6.1 INTRODUCTION

We say $f: X \times Y \to R$, $X \subseteq R^n$, $Y \subseteq R^m$ is a convex-concave function, if $\forall y \in Y$, $f(\cdot,y)$ is a convex function, and $\forall x \in X$, $f(x,\cdot)$ is a concave function. Convex-concave functions and concave-convex functions are also called saddle functions ([R]).

We say a point (x^*,y^*) is a saddle point for convex-concave function $f(x,y)$ if

$$f(x^*,y) \le f(x^*,y^*) \le f(x,y^*), \ \forall x \in X, \ \forall y \in Y.$$

Equivalently,

$$\inf_{x \in X} \sup_{y \in Y} f(x,y) = f(x^*,y^*) = \sup_{y \in Y} \inf_{x \in X} f(x,y).$$

For a constrained convex minimization problem, one can associate a saddle function, i.e. the Lagrangian whose saddle point provides a solution to the (primal) convex minimization, and the corresponding dual (concave) maximization problem.

Existence of a saddle point is guaranteed by Kakutani's fixed point theorem [K1] when X, Y are compact convex sets. For various saddle point theorems see Stoer and Witzgall [S 40], and Rockafellar [R 7].

It is well-known that through Lagrangian theory, a linear program can be transformed to a bilinear convex-concave function.

Let us define $\phi(x)$, and $\psi(y)$ as

$$\phi(x) = \sup_{y \in Y} f(x,y)$$

$$\psi(y) = \inf_{x \in X} f(x,y).$$

It is well-known that $\phi(x)$ is convex, and $\psi(y)$ is concave. In general it is very difficult to evaluate $\phi(x)$ and $\psi(y)$, as well as sets $Y(x) = \{y\colon f(x,y) = \phi(x)\}$, $X(y) = \{x\colon f(x,y) = \psi(y)\}$. If we assume that an oracle computes $\phi(x)$, as well as a subgradient $g(x) \in \partial\phi(x)$, and similarly for $\psi(y)$, the results of this chapter follow from the main theorem of Judin-Remirovskii on the ellipsoidal algorithm [J 8], which is also presented in a previous chapter.

Gradient or more generally subgradient methods are suggested to solve saddle point problems. A typical algorithm is

$$x^+ = \pi_X(x - \gamma\partial_x)f(x,y))$$

$$y^+ = \pi_Y(y + \gamma\partial_y)f(x,y)) \ ,$$

where $\pi_A(x)$ represents the projection of x onto set A, $\partial_x f(x,y)$ is the subgradient of the convex function $x \to f(x,y)$, and similarly $-\partial_y f(x,y)$ is the subgradient of $y \to -f(x,y) \cdot \gamma$ is the step length.

Under suitable regularity conditions, for $\gamma_k \to 0$, $\Sigma\gamma_k = \infty$, we have $\lim_k \rho(x_k,X^*) \to 0$, $\lim \rho(y_k,Y^*) \to 0$ where X^*, Y^* are set of corresponding saddle paints, i.e. $x^* \in X^*$, $y^* \in Y^*$ then (x^*,y^*) is a saddle point, and $\rho(x^*,X^*)$ is the Hausdorff metric (see, Golhstein [G 43], Maistrovskii [M 2], Demjanov [D 14-18], Korpelevich [K10], Poljak [P 15]).

6.2 AN ELLIPSOIDAL APPROACH TO CONVEX-CONCAVE FUNCTION OPTIMIZATION

Here we directly deal with $f(x,y)$ without trying to evaluate $\phi(x)$ and $\psi(y)$, and using only subgradients of $f(x,y)$.

We may restrict ourselves to compact $X \subseteq R^{n_1}$, $Y \subseteq R^{n_2}$. Thus we may assume that $\exists R_o > 0$ such that all such $(x^*,y^*) \in S_o \equiv \{z \in R^n : ||z|| \le R_o\}$, $n = n_1 + n_2$.

Let us represent $z = (x,y) \in R^{n_1} \times R^{n_2} = R^n$. We will denote $g_f^x(x)$ a subgradient of the function $x \to f(x,y)$ for a fixed y, and similarly $-g_f^y(z)$ that of $y \to -f(x,y)$. Let us define $g(z) = (g_f^x(z), -g_f^y(z))$ a particular member of the subgradient set of f

$$g(z) \in G(z) = G_f^x(x,y) \times (-G_f^y(x,y))$$

Lemma 1: $(g(z), z - z^*) \ge 0$

Proof: [S19]

$$0 \le f(x,y^*) - f(x^*,y) = f(x,y^*) - f(x,y) + f(x,y) - f(x^*,y)$$
$$\le (-g_f^y(z), y - y^*) + (g_f^x(z), x - x^*) = (g(z), z - z^*)$$

□

We will describe the algorithm as in Chapter 3.

0. Choose $z_o \in R^n$, B_o, $h_o = R_o/(n+1)$, $k = 0$

$$\alpha = \frac{n}{\sqrt{n^2 - 1}}, \qquad \beta = \sqrt{\frac{n-1}{n+1}}$$

1. Do while $k \le K(\varepsilon)$, and $g_k \in G(z_k)$, $g_k \ne 0$

$$s_k = B_k^T g_k / ||B_k^T g_k||$$

$$z_{k+1} = z_k - h_k B_k s_k$$

$$B_{k+1} = B_k R_\beta(s_k)$$

$h_{k+1} = h_k\alpha$.

$k = k + 1$, go to 1.

Let $d = \sup_{z \in S_o} ||g(z)||$

and $K(\varepsilon) = 4n^2 \lceil \log d + \log R_o + \log 1/\varepsilon \rceil$

Theorem 2: The ellipsoidal algorithm will give a point $(\bar{x},\bar{y})$ such that

$$||g(\bar{x},\bar{y})|| \; ||(\bar{x},\bar{y}) - x^*,y^*)|| < \varepsilon$$

in most $K(\varepsilon)$ iterations.

Proof: See Theorem 3.7.9

Since

$$0 \le f(\bar{x},y^*) - f(x^*,\bar{y}) = (f(\bar{x},y^*) - f(\bar{x},\bar{y})) + (f(\bar{x},\bar{y}) - f(x^*,\bar{y})),$$

at least one of the bracketed terms is non-negative; say, the second one. Now, consider

$$0 \le f(\bar{x},\bar{y}) - f(x^*,\bar{y}) \le \left\langle g_f^x(x,y), \bar{x} - x^* \right\rangle,$$

from which it follows that

$$|f(\bar{x},\bar{y}) - f(x^*,\bar{y})| \le |\left\langle g_f^x(\bar{x},\bar{y}), \bar{x} - x^* \right\rangle|$$

$$\le ||g_f^x|| \; ||\bar{x} - x^*|| \le ||g(\bar{z})|| \; ||\bar{z} - z^*|| \le \varepsilon.$$

Furthermore by the saddle point property $f(\bar{x},y^*) - f(x^*,y^*) \ge 0$ and $f(x^*,y^*) - f(x^*,\bar{y}) \ge 0$, hence

$$0 \le f(\bar{x},y^*) - f(x^*,\bar{y}) = (f(\bar{x},y^*) - f(x^*,y^*)) + (f(x^*,y^*) - f(x^*,\bar{y}))$$

which is $\leq\varepsilon$ by Lemma 1 and Theorem 3. Thus each of the bracketed terms are $\leq\varepsilon$.

Since $0 \leq f(\bar{x},y^*) - f(x^*,\bar{y}) \leq \varepsilon \Rightarrow |f(x^*,\bar{y}) - f(\bar{x},y^*)| \leq \varepsilon$,

$$|f(\bar{x},\bar{y}) - f(x^*,y^*)| \leq |f(\bar{x},\bar{y}) - f(x^*,\bar{y})| + |f(x^*,\bar{y}) - f(x^*,y^*)| \leq \varepsilon + \varepsilon = 2\varepsilon$$

$$\therefore \quad |f(\bar{x},\bar{y}) - f(x^*,y^*)| < 2\varepsilon$$

where $(\bar{x},\bar{y}) = (x_{\ell o},y_{\ell o}) = \operatorname{argmin} \{||B_\ell^t\, g(x_\ell,y_\ell)||: \ell \leq K(\varepsilon)\}$.

Thus we have arrive at the following

<u>Definition</u>: The pair $(\bar{x},\bar{y})$ is ε-optimal for $f(x,y)$ if $f(\bar{x},\bar{y}) - f(x^*,y^*)| \leq \varepsilon$ where (x^*,y^*) is a saddle point of f.

Letting $\varepsilon' = \varepsilon/2$ we have

<u>Corollary 3</u>: The ellipsoidal algorithm gives an ε-optimal solution in $K(\varepsilon')$ iterations for the above convex-concave game.

6.3 CONVEX-CONCAVE GAMES OVER SIMPLECES

A convex-concave game is a zero-sum game, in which the pay-off to the second player is defined by a convex-concave function $f(x,y): X \times Y \to R$. To solve the game is equivalent to finding a saddle point of f.

By restricting X and Y to be polyhedral sets, and modifying the algorithm presented above, we will obtain a polynomial algorithm for finding an ε-optimal solution to the above problem.

Let us consider the case where X and Y are simpleces, i.e.

$$X = \{x \in R^{n_1} ; x \geq 0 \quad , \quad \sum_{i=1}^{n_1} x_i = 1\}$$

$$Y = \{y \in R^{n_2} ; y \geq 0 \quad , \quad \sum_{i=1}^{n_2} y_i = 1\} .$$

Let $H = \{(x,y) \in R^{n_1+n_2} : \langle e_{n_1}, x \rangle = 1, \ \langle e_{n_2}, y \rangle = 1\}$

where e_n is the vector of 1's of dimension n.

For $x \in X$, $y \in Y$, we have $||x|| \leq 1$, $||y|| \leq 1$.

Let $x^o = \frac{1}{n_1} e_1$, $y^o = \frac{1}{n_2} e_2$, then $(x^o, y^o) \in H$.

We will restrict our ellipsoids to $(z,B) \subset H$. The basic idea is to keep the center of the ellipsoid feasible, i.e. $z \in R^n_+ \cap H$, and still preserve the convergence rate of the basic algorithm. Our motivation came from works of Poljak et.al (see [P 15]). A similar approach is employed by Judin-Nemirovskii [N 2]. Let

$$P_1 = I_{n_1} - \frac{e_{n_1} e^t_{n_1}}{n_1} , \qquad P_2 = I_{n_2} - \frac{e_{n_2} e^t_{n_2}}{n_2}$$

and

$$P_{\bar{H}} = P_1 \oplus P_2 , \text{ where } \bar{H} = H - (x^o, y^o) .$$

If

$$E = \begin{bmatrix} e^t_{n_1} , & 0 \\ 0 & e^t_{n_2} \end{bmatrix} ,$$

then

$$E \, P_{\bar{H}} \binom{x}{y} = 0, \ \forall (x,y) \in R^n .$$

In a typical iteration of the ellipsoidal algorithm, we have

$$z^+ = z - BB^t g$$

$$B^+ = BR_\beta(s_k) .$$

We initialize $B_o = P_{\bar{H}}$. Since $(z,B) \subset H$ implies that $z \in H$, and $E\,By = 0$, $\forall y \in R^n$, we have $z^+ \in H$, and $E\,B^+ y = 0$, $\forall y \in R^n$. Thus we have shown that $(z^+, B^+) \subset H$.

Lemma 4: Let (z,B) be an ellipsoid, $Q \subset (z,B)$ be a convex set, and $z \notin Q$. Let $\pi(z)$ be the projection of z onto Q. Then $Q \subset (\pi(z),B)$.

Proof: Without loss of generality we may assume (z,B) is a sphere. The lemma follows from the properties of projection.

We keep $(z,B) \subset H$ by choosing the initial ellipsoid suitably. In order to satisfy $z \geq 0$, we let $Q = H \cap R^n_+$ in Lemma 4, and we use the lemma. We need to show that projection can be done in polynomial time.

Since the effective dimension of the problem (dim H) is n-2, we should adjust the parameters β, r, $h = R_o/n+1$, by changing n to n-2. The same goes for n in $K(\varepsilon)$.

In fact, $$R_o^2 = \frac{n_1-1}{n_1} + \frac{n_2-1}{n_2} < 2.$$

If, in addition f is Lipschitz with M, then $K(\varepsilon)$ becomes

$$K(\varepsilon) = 4(n-2)^2 \log \frac{2\cdot M}{\varepsilon} ,$$

$$K(\varepsilon) = O(n^2 \log M/\varepsilon).$$

One can apply the above algorithm when X and Y are polyhedral sets. Then if ((x,y),B) is the current ellipsoid, and if $x \notin X$, and/or $y \notin Y$, let us define $\pi(x)$ as the projection of x onto X, (similarly $\pi(x)$). Then one can take $((\pi(x),\pi(y)),B)$ as the new ellipsoid, evaluate g(z), $z = (\pi(x),\pi(y))$ and continue the algorithm. Projection on a polyhedron is an instance of quadratic programming, for which a polynomial algorithm exists; namely the ellipsoidal algorithm [K 22,C22]. Thus one can show the existence of a polynomial algorithm for convex-concave games over polyhedral sets.

One should note that the projection in Lemma 4 is the non-orthogonal projection defined by the current ellipsoid. One expects that working with the defining inequalities would be faster.

7 Polyhedra theory and combinatorial optimization

In this chapter we will review the impact of Ellipsoidal Algorithm on polyhedra theory and combinatorial optimization. For the sake of being complete, we allow certain repetations.

Main references for this chapter are Grötchel-Lovász-Schrijver [G48,G49], Lovász-Schrijver [L13], Lovász [L15], Bachem and Grötchel [B 2].

7.1 POLYHEDRA THEORY

Given $A \in R^{m\times n}$, $b \in R^m$ the set

$$P(A,b) = \{x \in R^n |\ Ax \le b\} \tag{1}$$

is a polyhedron defined by A, b, and conversely any set $P \subseteq R^n$ for which $\exists\, A \in R^{m\times n}$, $b \in R^m$ such that $P = P(A,b)$ is a polyhedron. In other words, a polyhedron is the solution set of a finite system of inequalities. Sometimes, we will write P(A,b) as

$$P = \{x |\ a_i x \le b_i,\ i \in I\}. \tag{2}$$

$x \in P$ is a vertex of P if it is the unique solution $a_i x = b_i$, $i \in I'$, for some $I' \subseteq I$. For $x \in P$, there is a natural partition of I, as $I = I_1 \cup I_2$, where

$$I_1 = I_1(x) = \{i\colon a_i x = b_i\}, \tag{3}$$

$$I_2 = I_2(x) = \{i\colon a_i x < b_i\}. \tag{4}$$

Then $x \in P$, is a vertex of P if $r(A_{I_1})$, the rank of the subsystem induced by I_1 is n. It is well-known that a vertex is an <u>extreme point</u> of P, when P is treated as a convex set, i.e. it can not be written as a convex combination of other points in P. Let $I' \subset I$, $|I'| = n$, $r(A_{I'}) = n$, then the solution of $a_i x = b_i$, $i \in I'$ is a <u>basic feasible solution</u> if $x \in P$. For the system

$$P = (A,b) = \{x: Ax = b,\ x \geq 0\}\ , \tag{5}$$

the above definition reduces to the following: for some column-wise partition of A as $A = (A_1, A_2)$ and corresponding partition of $x = (x_1, x_2)$, $x \in P$ is a basic feasible solution of (5) if

$$r(A_1) = m, \text{ and } x_2 = 0, \tag{6}$$

where $m \leq n$ in this case. It is well-known that every basic feasible solution is a vertex, hence an extreme point of the underlying polyhedron. In general for a given vertex, we may have several basic feasible solutions. It is well-known that (1), (2), and (5) are equivalent in the sense that each can be written in other forms, if necessary, by introducing slack variables, replacing one variable by difference of two non-negative variables, etc.

Dimension of P, dim P, is the linear dimension of affine hull of P, aff (P). In other words, it is one less than the affine dimension of P. In tern, the latter is equal to the number of vertices in any maximal simplex that can be placed in P.

One fundamental result in the theory of linear inequalities is Farkas' lemma.

<u>Lemma 1</u>: Only one of the following two systems has a solution

i) $Ax \leq b$ (7)

ii) $yA = 0, \quad yb < 0, \quad y \geq 0.$ (8)

□

$\alpha x \leq \beta$ is called a valid inequality for P(a,b) if

$$\forall x \in P(A,b) \Longrightarrow \alpha x \le \beta \ . \tag{9}$$

An equivalent statement of Farkas' lemma is the following:

<u>Lemma 2</u>: $\alpha x \le \beta$ is valid for $P(A,b) \Longleftrightarrow \exists\, y \ge 0$, s.t.

$$yA = \alpha, \quad yb \le \beta \ . \tag{10}$$

□

Proof is based on the first finite algorithm for linear inequalities: Fourier elimination. For a modern treatment of Fourier method and a proof of lemma, see e.g. Bachem and Grötchel [B 2], Stoer and Witzgall [S40].

Thus, Farkas' lemma gives a characterization of non-emptiness of a polyhedron. It is well-known that the simplex method also gives a constructive proof of Farkas' lemma; and conversely main duality theorem of linear programming can be proved by Farkas' lemma.

Let us consider the linear program

$$\begin{aligned} &\max \ cx \\ \text{s.t.}\ & \\ &Ax \le b. \end{aligned} \tag{11}$$

Its dual is

$$\begin{aligned} &\min \ yb \\ \text{s.t.}\ & \\ &yA = c \\ &\ y \ge 0. \end{aligned} \tag{12}$$

The following is the main Duality Theorem:

<u>Theorem 3</u>: For the pair of LP's (11) and (12) exactly one of the following holds:

i) $\exists\ \bar{x}$ solving (11), and $\bar{y}$ solving (12). Furthermore $c\bar{x} = \bar{y}b$.

ii) One of (11), (12) is unbounded, and the other is infeasible.

iii) Both of (11), (12) are infeasible.

□

It is well-known that one can use main duality theorem to prove various theorem of alternatives (see, e.g. Dantzig [D 6], Mangasarian [M 6] .

For the system $Ax \le b$, $x \ge 0$ let $\ell_1, \ell_2, \ldots, \ell_n$ be norms of columns of A, and let D be multiplication of m largest of them. Then, it is easy to verify via Hadamard theorem and Cramer's rule that for any basic solution of $[A,I]x = b$, $x \ge 0$ we have

$$\left. \begin{array}{l} |x_i| \le D, \\ \text{and} \\ x_i \ne 0 \Longrightarrow |x_i| \ge \frac{1}{D} . \end{array} \right\} \qquad (13)$$

Phase I of the simplex method provides a constructive finite proof of Farkas' lemma. Simply apply phase I, either algorithm will give us a feasible solution of $Ax = b$, $x \ge 0$, or show its non-existence and provide a $y \ge 0$ s.t. $yA = 0$, $yb < 0$.

The ellipsoidal algorithm uses a characterization of non-solvability of a linear inequality system in a crucial way. In Chapter 3 (3.8) we have provided several proofs of the following

<u>Theorem 4</u>: If $P(A,b)$ is empty then $\theta(x) \ge 2 \cdot 2^{-L}$ □ (14)

Here, $\theta(x)$ is defined as

$$\theta(x) = \max_i \{a_i x - b_i\} .$$

Now, we will give a proof of (14) using the simplex method. Apply phase I to system $Ax \le b$. Either one can introduce the constraints $x \ge 0$ by known techniques and write the system in standard equality form, or one can work with semi-linear programming method of Fourer [F8]. Anyway, in both cases the same A will work. When phase I stops, the sum of the values of artificial variables is equal to $\theta(x^*)$. Since by assumption the original linear inequality system is infeasible, at least one artificial variable will remain in the basis with a positive value. Clearly, each positive value on the right hand side $\geq \frac{1}{D}$, proving $\theta(x^*) \geq \frac{1}{D} > 2 \cdot 2^{-L}$.

Gács-Lovász [G13] in their exposition of the ellipsoidal algorithm, as described earlier, studied a different problem. They considered the system

$$Ax < b. \tag{15}$$

Note that $\bar{P}(A,b) = \{x: Ax < b\}$ is an open set. If it is non-empty then it is full-dimensional, since one can place a ball inside the solution set. They proved the following

<u>Theorem 5</u>: If $\bar{P} \neq \phi$, then volume of $\bar{P}$,

$$\text{vol}(\bar{P}) \geq 2^{-(n+1)L} . \qquad \square$$

They proved this by exhibiting an n-simplex within $\bar{P}$ with rational coordinates and calculating its volume.

Here we will give a different proof. Define $f_i = \lceil ||a_i|| \rceil$, where a_i is the i'th row of A, and $\lceil x \rceil$ is the smallest integer $\geq x$. Consider the linear program

$$\begin{aligned} &\max t \\ &a_i x + f_i t \leq b_i, \quad i \in I \\ &\qquad t \geq 0 . \end{aligned} \tag{16}$$

By hypothesis, the system has a solution with $t > 0$. Let $(\bar{x}, \bar{t})$ be an optimal solution of (16). Define

$$I' = \{i \in I: a_i \bar{x} + f_i t = b_i\} . \tag{17}$$

Clearly

$$\bar{t} \geq \{|\det(A_{I'}, f_{I'})|\}^{-1} . \tag{18}$$

Let

$$a_i' = (a_i, f_i) \in R^{n+1} .$$

Then

$$|\det(a_i', \ i \in I')| \leq \prod_{i\in I'} ||a_i'|| \leq 2^{n/2} \prod_{i\in I'} ||a_i|| < 2^L . \tag{19}$$

Thus

$$\bar{t} \geq 2^{-L} . \tag{20}$$

Clearly the sphere centered at $\bar{x}$ with radius 2^{-L} is contained in $\bar{P}$. The volume of this sphere is

$$\omega_n \ 2^{-nL} ,$$

where ω_n is the volume of the unit sphere in R^n . Thus

$$\mathrm{vol}(\bar{P}) \geq \omega_n \ 2^{-nL} . \qquad \square \tag{21}$$

This automatically gives an upper bound for the solution of strict linear inequality system; $K = 4n^2L$. Recall that the volume of the initial ellipsoid is $\omega_n \ 2^{nL}$ and rate of volume reduction is

$$e^{-1/2n} < 2^{-1/2n} .$$

Gács-Lovász used the following theorem to solve P via $\bar{P}$.

<u>Theorem 6</u>: The system

$$a_i x < b_i + 2^{-L} , \ \forall i \tag{22}$$

has a solution $\Longleftrightarrow$ the system

$$a_i \leq b_i , \ \forall i \tag{23}$$

has a solution. $\square$

A careful look at the proof in [G13] or proof of Theorem 4.2 will show that the following is true:

Theorem 7: Let $x(\varepsilon)$ be a vertex of

$$a_i x \le b_i + \varepsilon \ , \ \forall i \tag{24}$$

for $0 < \varepsilon \le 2^{-L}$, i.e. $x(\varepsilon)$ is the unique solution of

$$a_i x = b_i + \varepsilon \ , \ i \in I' \subset I$$

then, solution $\bar{x}$ defined by

$$a_i x = b_i \ , \ i \in I'$$

is a vertex of (25)

$$a_i x \le b_i \ , \ i \in I \qquad \square$$

Remark: The bound on ε can be slightly increased. For example $\varepsilon = 1/nD$, where D is the largest determinant among the row basis of A will work.

In polyhedra theory, there is a process known as "pulling vertices", which allows every vertex to be on exactly n hyperplanes (facets). In case of degeneracy one may have more than n hyperplanes at a vertex. In that case for a single vertex one has several basic feasible solutions. "Pulling vertices" puts basic feasible solutions and vertices into 1-1 correspondence (see Graünbaum [G50], Lee [L 7]). Thus as a byproduct of ellipsoidal algorithm, one obtains an explicit upper bound on the perturbation of the right hand side for this purpose.

A similar method suggested by Charnes [C10] in order to prevent cycling in simplex method. One would replace the right hand side with

$$b(\varepsilon) = b + \Sigma e_i \varepsilon^i \ . \tag{26}$$

In actual implementation of the algorithm, one does not need the actual value of ε. Again as a byproduct of ellipsoidal algorithm we have an a priori upper bound on ε. Note that we can not take ε arbitrarily large, since a vertex of the perturbed polyhedron may correspond to a non-feasible basic point of the original polyhedron.

Goffin [G37] studied effects of such a perturbation on some lattices associated with polyhedra.

Since by virtue of the ellipsoidal algorithm linear programming is polynomial, any algorithm which uses linear programs as a subroutine in a polynomial number of times becomes polynomial. One such instance is the identification of redundant and binary constraints, and obtaining a minimal representation of a polyhedron.

Suppose we are given

$$P = \{x: a_i x \le b_i \ , \quad i \in I\} \ne \phi \ ,$$

we would like to determine whether $P = P'$ where

$$P' = \{x: a_i x \le b_i \ , \quad i \in I, \quad i \ne 1\} . \tag{27}$$

It suffices to solve the following LP

$$\begin{aligned} &\max \quad a_1 x \\ &\text{s.t. } x \in P' \ . \end{aligned} \tag{28}$$

Let x^* be an optimal solution. If $a_1 x^* < b_1$ we known that the equation

$$a_1 x = b_1 \tag{29}$$

is never satisfied, and hence can be dropped. This constraint is called <u>redundant</u>. If on the other hand $a_1 x^* \ge b_1$, it is irredundant. When $a_1 x^* = b_1$, if we solve

$$\begin{aligned} &\min \quad a_1 x \\ &x \in P' \end{aligned} \tag{30}$$

with x^* as an optimal solution. If $a_1 x^* = b_1$, then the constraint

(29) is <u>binding</u>, or <u>implicit equality</u>.

A <u>minimal representation</u> of $\{x: a_i x \le b_i, i \in I\}$ is a subsystem with no redundant constraints. By solving problems of the type (28) and (30) at most $2|I|$ times, one can find a minimal representation of a given polyhedron, and identify the binding constraints. Note that in general a minimal representation is not unique, and depends on the original ordering of the constraints.

Now consider the standard LP

$$\begin{aligned} \max \quad & f(x) = cx \\ \text{s.t.} \quad & Ax = b \\ & x \ge 0 \ . \end{aligned} \tag{31}$$

Let $D \in Z_+$ be such that for every basic solution satisfies (13). If $\bar{x}$, $\tilde{x}$ are two different basic (feasible) solutions with $f(\bar{x}) \neq f(\tilde{x})$, it is clear that

$$|f(\bar{x}) - f(x)| \ge \frac{1}{D} \ . \tag{32}$$

Recall that such an observation is used in Chapters 3, 4 and 5. One can perturb c so that new (31) has a unique solution, if indeed it has a solution. Let $\varepsilon > 0$, be a rational number so that

$$D \sum_{j=1}^{n} \varepsilon^j \le \frac{1}{2D} \ . \tag{32}$$

Since

$$D \sum_{j=1}^{n} \varepsilon^j \le nD\varepsilon \ , \text{ for } \varepsilon < 1 \ ;$$

$$\varepsilon = \frac{1}{2nD^2}$$

will be sufficient. Now, let c' be

$$c' = c + (\varepsilon, \varepsilon^2, \ldots, \varepsilon^n). \tag{33}$$

Clearly for $f'(x) = c'x$, $\max\{f'(x) | Ax = b, x \geq 0\}$ will have a unique solution; namely the lexicographic maximum of the solutions of (31). By multiplying c' by $(1/\varepsilon)^n = (2nD^2)^n$, we obtain a new objective function for which (31) has a unique solution, which also solves (31) with the old objective function.

Given $P = \{x \in R^n : a_i x \leq b_i, i \in I\}$, and $y \in P$, one is often interested in an explicit representation of y as a convex combination of vertices of P, assuming P is bounded. In other words, we would like to determine vertices $x_o, x_1, \ldots, x_n$ such that

$$\begin{aligned} y &= \sum_{i=0}^{n} \lambda_i x_i \\ 1 &= \Sigma\lambda_i \\ \lambda_i &\geq 0. \end{aligned} \tag{34}$$

Existence of such a representation is guaranteed by polyhedral decomposition theorem and carathéodory theorem. Grötchel-Lovász-Schrijver gave a polynomial algorithm using the ellipsoidal method. In fact, they did not assume an explicit representation of P as a system of inequalities: they required a separation algorithm. Some of their results will be discussed later. One can implement their ideas in polynomial time using simplex method or projection method. Recall our algorithm VERTEX (see Chapter 2); given P, $y \in P$ and c, will find a vertex $\bar{x} \in P$, with $c\bar{x} \geq cy$ in n simplex iterations. The algorithm constructs a sequence $x_o, x_1, \ldots, x_n$ of vertices of P, $y_o, y_1, \ldots, y_n \in P$, and $F_1, F_2, \ldots, F_n$ facets of P. Let x_o be the vertex obtained by VERTEX starting with y, and $c = 0$. Let $y_o = y$. Assume x_i, y_i, F_i are defined for $i \leq j$. Let

$$y_{j+1} = x_j + t(y_j - x_j), \tag{35}$$

for largest $t \in R$ such that $y_{j+1} \in P$. Let $j \leftarrow j + 1$. Then

$y_j \in \{x \in P: a_j x = b_j\}$ and $a_j(y_{j-1} - x_{j-1}) > 0$. Obviously $a_j x = b_j$ is a facet of P, say F_j. Then x_j is the vertex obtained from y_j by VERTEX with the objective function $c = \sum_{i=1}^{j} a_i$. Clearly $x_i, y_i \in F_j$, for $j \geq i$, and $y \in co(x_o, .., x_i, y_i)$, $\dim(F_1 \cap, ..., \cap F_i) = n-i$. Hence $x_n = y_n$, so $y \in co(x_o, x_1, .., x_n)$. The determination of λ is a simple task. The unique solution of

$$y = \sum_{i=0}^{n} \lambda_i x_i$$

$$1 = \Sigma \lambda_i$$

is the required λ. Thus we have proved that

Theorem 8: Given $y \in P$, a representation of y as a convex combination of vertices of P can be found in (n+1) application of VERTEX, when P is bounded. □

When P is not bounded and if y has a component, say z, in the recession cone, then the algorithm VERTEX will detect it through (35). In that case, by intersecting recession cone with a hyperplane in "general position" one can identify the generators of the recession cone with the vertices of the polytope generated by this intersection. By applying the above algorithm to the new polytope one obtains a representation of z in terms of extreme rays of P. For details see [A14].

7.2. COMBINATORIAL OPTIMIZATION

A typical combinatorial optimization problem is the following. Given a finite set F find

$$\max \{cx \mid x \in F\}, \tag{1}$$

(here we are identifying $x \in F$, with its incidence vector). Usually F is exponentially large in n, the number of components of x. Let

$$P = \text{co}\ \{x: x \in F\}. \tag{2}$$

Then clearly

$$\max \{cx \mid x \in F\} = \max \{cx \mid x \in P\}. \tag{3}$$

It is well-known that P is a convex polyhedron. It is also well-known that for any polyhedron P, there exists A,b such that

$$P = \{x: Ax \le b\}. \tag{4}$$

Depending on structure of F, the number of rows of A can be exponential in n. So there are 2 main difficulties in solving (1) as a linear program. First we do not have a readily available representation of P as (4), i.e. an explicit knowledge of A and b. Secondly, we do not have any polynomial algorithm to solve any LP, other then the ellipsoidal algorithm and recent Yamnitsky-Levin algorithm (see Chapter 8). In very few cases we have an explicit description of the underlying polyhedron for a combinatorial optimization problem: b-matchings ([E 7], branchings [E 6 ,F 13], (poly)-matroid [Z16], (poly)-matroid intersection [Z17], perfect graphs [L12,C14], etc. They all depend on very deep results on the structure of F.

Let us note once more that the ellipsoidal algorithm, (and Yamnitsky-Levin algorithm) solves

$$\max \; cx \qquad Ax \le b \tag{5}$$

with an upper bound polynomial in n, and some problem data. If we have any other algorithm which solves

$$\max \; cx \qquad Ax = b \qquad x \ge 0 \tag{6}$$

with a complexity as a polynomial function of n, it is not acceptable in combinatorial optimization; since the transformation from (5) to (6) will move n to m+n, which is exponential in n. In other words the above transformation is not a polynomial transformation for combinatorial problems.

Following Grötchel-Lovász-Schrijver, let us introduce some formulation. Let K be a non-empty compact, convex set in R^n .

SOP: Strong Optimization Problem. Given $c \in R^n$ solve $\max \{cx | x \in K\}$.

SSP: Strong Separation Problem. Given $y \in R^n$ decide whether $y \in K$, and if not find a separating hyperplane, i.e. find $c \in R^n$ s.t. $cy > \max \{cx | x \in K\}$.

When K is given as $K = \{x: Ax \le b\}$, the separation problem is trivial, where as optimization problem is the difficult one. But, if K is given as $K = co\{x_1,\dots,x_k\}$ for some $x_1,\dots,x_k \in R^n$, then optimization problem is trivial, whereas separation problem becomes difficult.

Even though strong optimization and strong separation problems are natural ones we have to pass weaker and complicated versions for technical reasons. Let

$$d(x,K) = \inf \{d(x,y) | y \in K\} \tag{7}$$

where d(x,y) is the euclidean distance between x and y.

WOP: Weak Optimization Problem - Given $c \in Q^n$, $\varepsilon > 0$ find

$$y \in Q^n \quad \text{s.t.} \quad d(y,K) \le \varepsilon, \quad cy \ge cx - \varepsilon, \quad \forall x \in K \tag{8}$$

WSP: Weak Separation Problem - Given $y \in Q^n$, $\varepsilon > 0$ determine either

i) $d(y,K) \le \varepsilon$, or

ii) find $c \in Q^n$, $||c|| \ge 1$, $c^T y \ge cx - \varepsilon$, $\forall x \in K$. (9)

Moreover, Grötchel-Lovász-Schrijver made the following technical assumption: $\exists\, a_o \in Q^n$, $0 < r \le R$ such that

$$S(a_o,r) \subseteq K \subseteq S(a_o,R) \tag{10}$$

where $S(x,t)$ denotes the euclidean ball of radius t at center x. This means that K is a full-dimensional, compact convex set. Theoretically for a nontrivial compact K this is possible; one has to work in the minimal affine flat containing K. But for a given K to find such a description is a formidable task.

Grötchel-Lovász-Schrijver proved the following

Theorem 1: Let K satisfy (10), and let $0 < \varepsilon < r \le R$. Assume there is a subroutine SEP to solve the (weak) separation problem for K. Then, the ellipsoidal algorithm will solve (weak) optimization problem in $O(n^2 \log 2R^2||c||/r\varepsilon)$ iterations. □

The problem we are trying to solve is $\max\{cx \mid x \in K\}$, which can be written as

$$\begin{aligned} &\max \quad cx \\ &\text{s.t.} \quad d(x,K) \le 0. \end{aligned} \tag{11}$$

Since K is convex, $d(\cdot,K)$ is a convex function on R^n. Thus results of Judin-Nemirovskii [J8] applies to this case. In fact one can imitate even Hachiyan's proof of the main result (see 3. 9), and obtains the same theorem. Let

$$M_\varepsilon(x) = \begin{cases} -c & \text{if} \quad d(x,K) \le \varepsilon \\ \partial d(x,K) & \text{if} \quad d(x,K) > \varepsilon, \end{cases} \tag{12}$$

then one chooses

$$g(x) \in M_\varepsilon(x) \tag{13}$$

at each iteration of the ellipsoidal algorithm. In other words, when the point $x \in K + \varepsilon B$, where B is the unit ball, we take the objective function cut, otherwise we take any subgradient of $d(x,K)$ at x. The latter can be substituted by any separating hyperplane of K at x. $\partial(\cdot,K)$ is the normal cone of K in the Clarke's sense when K is convex [C 24]. The new ellipsoid contains the half of the old ellipsoid defined by the separating hyperplane. Now, let

$$N = N(\varepsilon) = 2n^2 \left\lceil \log\left(\frac{R||c||}{\varepsilon}\right) \right\rceil, \tag{14}$$

in $N(\varepsilon)$ iterations the volume of the current ellipsoid will reduce (upto a factor of ω_n)

$$R^n e^{-\frac{2n^2}{2n} \log \frac{R||c||}{\varepsilon}} = R^n \left(\frac{\varepsilon}{R||c||}\right)^n = \frac{\varepsilon^n}{||c||^n}, \tag{15}$$

so that the inradius (thickness) of the ellipsoid E_N

$$r(E_N) \le \frac{\varepsilon}{||c||}. \tag{16}$$

Now, consider the ellipsoid at iteration N, we know that $x^* \in E_N$. The ball centered at x^*, with radius $\varepsilon/||c||$ will hit one of the hyperplanes generated during the algorithm. Let us say, it hits the hyperplane indexed by k, at the point $\bar{x}_k$, and let x_k be the center of the ellipsoid E_k. Then we have

$$||x^* - \bar{x}_k|| \leq \frac{\varepsilon}{||c||} , \tag{17}$$

implying

$$|c(x^* - \bar{x}_k)| \leq \varepsilon . \tag{18}$$

Since H_k is a separating hyperplane, we have $d(x_k, K) \leq \varepsilon$, otherwise we would have a contradiction to $||\bar{x}_k - x^*|| \leq$, since $\bar{x}_k \in H_k$. (This is quite easy to see when the underlying convex set is a polytope. In general, it suffices to take the normal of the hyperplane by as $\tilde{x} - x$, where $d(x,\tilde{x}) = d(x,K)$.) Thus $c\bar{x}_k = cx_k$, and

$$f(x_{\ell_o}) = \max_{\ell \leq N} \{f(x_\ell) | d(x_\ell, K) \leq \varepsilon\} \geq f(x_k) = f(\bar{x}_k) \geq f(x^*) - \varepsilon, \tag{19}$$

implying x_{ℓ_o} is the ε-optimal solution. □

Thus we have proved

<u>Theorem 2</u>: Let K be non-empty, convex, compact set satisfying $K \subseteq S(0,R)$. Suppose we have a separation algorithm SEP. If the cuts are choosen according to (12) and (13), then the ellipsoidal algorithm will find an ε-optimal solution in $N(\varepsilon)$ iterations. □

Note that we are implicitly assuming that norms of the cuts used during the algorithm $\leq ||c||$.

Let us point out that K in Theorem 2 need not be full-dimensional. Consequently, we do not need an explicit knowledge of r. However, a perturbation of K is necessary in either case.

The above results are valid for a general convex, compact set K. When K is known to be a rational polytope, i.e. it is bounded and it comes from a linear inequality system with rational vertices, we can do more. By choosing ε sufficiently small, and with some additional work, which is also polynomial, we can solve the strong optimization problem.

Let us assume that K is a rational polytope, say P, and satisfies 1.13, i.e. $\exists\, D \in Z_+$ such that every basic point x,

satisfies $|x_i| \le D$, and $x_i \ne 0$, $|x_i| \ge \frac{1}{D}$. A well-known result in the theory of continued fractions states that if x is a rational number of the above form, y is a real number such that $||y-x|| \le \frac{1}{2D^2}$, rounding y to next rational number with denominator at most D will result in x, see, e.g. Hinchin [H16], Reiss [R 3]. Thus it suffices to choose ε as $\frac{1}{2\,D^2||c||}$.
This will guarantee that x_{ℓ_o} will give x^*, and $f(x_{\ell_o})$ will give $f(x^*)$.

Now, let us see some applications of these results in combinatorial optimization.

<u>Optimum b-matching problem</u>

Let $G = (V,E)$ be a graph, where V is the set of nodes, and E is the set of edges. Furthermore, let $b \in Z_+$ is given.
A b-matching $x = (x_j,\ j \in E)$ is an assingment of non-negative integers x_j to edge j, such that

$$\Sigma\{x_i:\ i \wedge v\} \le b_v\ ,\ \forall v \in V, \tag{20}$$

where the above summation over all edges incident with node v. Optimum b-matching problem is the selection of b-matching maximizing given objective function cx, $c \in Z^E$. Edmonds [E 3 , E 7] derived a description of the convex hull of all b-matching for a G. Thus, polytope of all b-matchings is given as

$$\begin{array}{ll} \text{i)} & x_i \ge 0\ ,\quad i \in E \\ \text{ii)} & \Sigma\{x_i:\ i \wedge v\} \le b_v\ ,\ v \in V \\ \text{iii)} & \Sigma\{x_i:\ i \in \delta(V')\} \ge 1\ ,\quad V' \in \mathcal{F} \end{array} \tag{21}$$

where $\delta(V')$ is the set of all edges with only one node in V', and $V' \in \mathcal{F}$ if $b(V') = \Sigma\{b_v:\ v \in V'\}$ is odd.

Thus b-matching problem can be cast as

$$\begin{array}{ll} \max & cx \\ \text{s.t.} & (21). \end{array} \tag{22}$$

By Theorem 1 or 2, problem is reduced to checking whether any $x \in R^E$ satisfies (21) or not. Clearly i) and ii) are very easy to verify. If they are not satisfied, then the cut is readily available. Padberg-Rao [P2] recently gave a polynomial algorithm to check iii) extending earlier work of Gomory-Hu. Padberg-Rao algorithm minimizes the summation on the left hand side of iii) over F. If the minimum ≥ 1, then x satisfies (21), otherwise, for minimizer V', iii) will give a valid cut.

Let us remind the reader that the blossom algorithm, although very efficient, is not polynomial in b; it is pseudo-polynomial. There are other polynomial algorithm for optimum b-matching. Cunnigham-Marsh-Pulleyblank [C 40], see also Marsh [M 32], Rao-Truemper [R 15], Akgül [A12], Anstee [A17]. They all employ Edmonds-Karp [E 9] scaling one way or another. When x_j is restricted to $0 \leq x_j \leq 1$, then iii) of (21) should be changed, see Padberg-Rao [P 5] for details.

Branchings

Let G = (V,E) be a directed graph with specified node r, called the root. Each edge: $j \in E$, has a tail t(j), and head h(j). A branching T is a directed spanning tree rooted at r, i.e. every node $v \neq r$, has exactly one edge in the tree whose head is at v. A rooted cut E', is a subset of edges so that $E' = \delta^+(V')$ for some $V' \subseteq V$, $r \in V'$, where

$$\delta^+(V') = \{j: t(j) \in V'\} .$$

By definition each rooted cut E' intersects each branching, i.e. $|E' \cap T| \geq 1$.

By Branching theorem [E 6 ,F13], it is known that the convex hull of the characteristic vectors of all branchings rooted at r is given by

$$\begin{aligned} &\text{i)} \quad 0 \leq x_j \leq 1, \quad \forall j \in E \\ &\text{ii)} \quad \Sigma\{x_j : j \in E'\} \geq 1 , \quad \forall \text{ rooted cut } E'. \end{aligned} \tag{23}$$

Thus, given weights c_j, $j \in E$, the optimum branching problem can be cast as

$$\begin{aligned} &\max \ cx \\ &\text{s.t.} \ (23). \end{aligned} \tag{24}$$

By Theorem 1 or 2, it suffices to exhibit a polynomial separation algorithm for (23) in order to solve (24) in polynomial time. Clearly i) of (23) is easy to check. To check ii) of (23), $\forall s \in V - \{r\}$ we solve a network flow problem. Let r be source, s be sink node on the network G = (V,E) with edge capacities given by x. We can use any maximal flow algorithm to find the value of min cut, i.e. left side of ii) in (23). If $\forall s$, $\Sigma\{x_j : j \in E'\} \geq 1$ then x is feasible, otherwise for any s, violating this will give a valid cut. Thus the separation algorithm is obtained by applying a maximum flow algorithm in $|V| - 1$ times. Since the maximum flow algorithm is polynomial, see e.g. [P 7], the separating algorithm is polynomial, hence (24) can be solved in polynomial time via ellipsoidal algorithm.

Let us remind the reader that previous polynomial algorithms for branching are known; Edmonds [E 6], Fulkerson [Fl3], in fact the characterization used above is proved by Edmonds algorithm. There is even a polynomial algorithm for a generalized branching problem: branching systems i.e. several branchings rooted possibly different roots, Calvillo [C 1].

The relationship between branchings and rooted cuts is an instance of a general structure. Let V be a finite set, E a collection of subsets of V such that no member of E is contained in another member of E. Then E is called a <u>clutter</u>. The <u>blocker</u> of E, $B(E)$, is the set of minimal subsets of V that meet every member of E, i.e. $B \in B(E) \iff B \cap E \neq \phi$, $\forall E \in E$, and B is minimal with respect to this property. Clearly $B(E)$ is a clutter and $B(B(E)) = E$. Let $P = P(E)$ be the convex hull of the characteristic vectors in R^V of all subsets of V which contain a member of E. Then $x \in P$ satisfies

$$\begin{aligned} &\text{i)}\ \ 0 \le x_j \le 1\ , \qquad j \in V \\ &\text{ii)}\ \ \Sigma\{x_j : j \in V'\} \ge 1\ ,\quad V' \in B(\mathcal{E}) . \end{aligned} \tag{25}$$

If P is completely determined by (25), (as in branchings), then we say $\mathcal{E}$ has the Q_+-max flow-min cut property (Q_+-MFMCP). Fulkerson [F12] has shown that if $\mathcal{E}$ has the Q_+-MFMCP, then so is $B(\mathcal{E})$. By replacing $B(\mathcal{E})$ with $B(B(\mathcal{E})) = \mathcal{E}$ in ii) of (25); we see that $P(B(\mathcal{E}))$ has the same representation,

$$\text{ii)'}\ \ \Sigma\{x_j : j \in V'\} \ge 1,\quad V' \in \mathcal{E} . \tag{26}$$

If have an algorithm which solve for a given x,

$$\min\ \{x(V') : V' \in \mathcal{E}\}\ , \tag{27}$$

where $x(V') = \Sigma\{x_j : j \in V'\}$, we know how to solve separation problem, hence the optimization problem over $P(B(\mathcal{E}))$. Thus we have

<u>Theorem 3</u>: Let $\mathcal{E}$ be a clutter with Q_+-MFMCP. Assume further that there is a polynomial algorithm to solve $\min\{w(V') : V' \in \mathcal{E}\}$, for $w \in Z^V$, then there is a polynomial algorithm to solve $\min\{w(V') : V' \in B(\mathcal{E})\}$. □

One important application of the above result is the following. Let $G = (V,E)$ be a directed graph. A set $D \subseteq E$ is a <u>directed cut</u> if, there is $V' \subset V$ such that $D = \delta^+(V')$, and $\delta^-(V') = \{j \in E : h(j) \in V'\} = \phi$. Clearly set of minimal directed cuts form a clutter, say $\mathcal{E}$. Blocker $B(\mathcal{E})$ is the set of coverings of G. In other words E' is a covering if contraction of E' makes D strongly connected. It is known that $\mathcal{E}$ has the Q_+-MFMCP, [L17]. Thus in order to solve optimization problem over $P(\mathcal{E})$ or $P(B(\mathcal{E}))$, it suffices to provide polynomial algorithms for $P(B(\mathcal{E}))$ or $P(\mathcal{E})$. Hence in order to solve minimum weighted covering, we have to solve minimum directed cut problem. But that is easy. Either we apply max-flow min cut theorem for each pair of nodes. Dijkstra's shortest path algorithm also yields a polynomial algorithm for minimum directed cut problem, by appealing to Theorem 3. For previous solutions see [L17,K12,F10].

Polarity

Let P be a rational polytope in R^n satisfying (1.13). First let us assume P is full-dimensional. The inequality

$$\pi x \le \pi_o \;,\; \pi \in R^n \;,\; \pi_o \in R \tag{28}$$

is a facet of P if the above inequality is a valid inequality and there are n affinely independent vertices of P satisfying (28) as equality. If we assume $\underline{0}$ is in the interior of P, then we eliminate all simple facets $e_i x \ge 0$, where e_i is the usual unit vector. One can normalize (28) so that $\pi_o = \pm 1$. Then considering the equation

$$D\,\pi x^i = \pi_o D \;,\; i = 1,\dots,n \tag{29}$$

which has integral coefficients, and $\pi_o = \pm 1$, one can determine upper and lower bounds on π_i. If $x^1, x^2, \dots, x^n$ are linearly independent, then

$$\left| \det \begin{pmatrix} Dx^1 \\ \\ Dx^n \end{pmatrix} \right| \le (D^2 \sqrt{n})^n = T, \tag{30}$$

so that every component of π satisfies $|\pi_i| \le T$, $\pi_i \ne 0$, $|\pi_i| \ge 1/T$. If P is not full-dimensional or $x^1,\dots,x^n$ are linearly dependent, then a subsystem of (29) has full rank, and any basic solution of (29) will give upper and lower bounds for π. So the above T will be valid in general case. Padberg and Rao [P5] studied facet inequalities thoroughly.

Let us assume, again that P is full-dimensional and (10) holds. Then the polar P^* of P is

$$P^* = \{y \in R^n : xy \le 1, \;\; \forall x \in P\} \; . \tag{31}$$

Let $x^1, x^2, \dots, x^k$ be the vertices of P. Then P^* can be written as

$$P^* = \{y : x^i y \le 1, \; i = 1,2,\dots,k\} \; .$$

P^* represents the set of all valid inequalities for P. It is well-known that the extreme points of P^* are facets of P, and dually $x^i y \le 1$ is a facet of P^*. It is easy to check that

$$(P^*)^* = P,$$

and

$$S(0, \frac{1}{R}) \subseteq P^* \subseteq S(0, \frac{1}{r}) .$$

For $c \in Q^n$, let $\alpha = cx^* = \max\{cx | x \in P\}$. Define

$$\bar{c} = \frac{c}{\alpha} ,$$

Then clearly

$$\bar{c}x \le 1, \quad \forall x \in P \Longrightarrow \bar{c} \in P^*.$$

Then $\bar{c}$ can be written as

$$\bar{c} = \sum_{i=0}^{n} \bar{\lambda}_i v_i, \quad \Sigma\lambda_i = 1, \quad \bar{\lambda}_i \ge 0, \tag{32}$$

where v_i is a vertex of P^*. Since

$$\bar{c}x^* = 1, \quad v_i x^* \le 1,$$

together with (32), we have

$$v_i \bar{x} = 1 \quad \text{for} \quad \lambda_i > 0.$$

Multiplying by a suitable integer b_i

$$v_i x \le 1 \Longleftrightarrow a_i x \le b_i, \quad a_i \in Z^n, \quad b_i \in Z.$$

Let

$$\tilde{\lambda}_i = \bar{\lambda}_i / b_i, \text{ and } \lambda_i = \alpha\tilde{\lambda}_i .$$

Then we have

$$c = \Sigma\lambda_i a_i \ , \ \lambda_i \geq 0 \ , \ cx^* = \Sigma\lambda_i b_i \ .$$

Note that this result can be obtained very easily when P is defined by a system of inequalities with integer coefficients via LP duality. If P^* is given by a system of linear inequalities, we can apply theorem 1.8, otherwise we have to appeal to ellipsoidal algorithm, as done in Grötchel-Lovász-Schrijver. Thus we have

Theorem 4: Let P be a rational polytope. Then there exists a polynomial algorithm, given $c \in Q^n$, which provides facets $a_i x \leq b_i$, and rational $\lambda_i \geq 0$ such that

$$c = \sum_{i=1}^{n} \lambda_i a_i \ , \ \Sigma\lambda_i b_i = \max \{c^T x | x \in P\} \ . \qquad \Box$$

Now, we have the machinery to prove

Theorem 5: For convex bodies polynomial optimization algorithm implies polynomial seperation algorithm.

Proof: For simplicity, let us consider a rational polytope. Given $\bar{x} \in R^n$, we would like to decide whether $\bar{x} \in P$, otherwise we want $c \in R^n$, $c\bar{x} \geq \max\{cx | x \in P\}$. Let us solve

$$\max\{\bar{x}y | y \in P^*\}. \qquad (33)$$

If $\bar{x}y \leq 1$, then $\bar{x} \in P$, otherwise $c = y$ is the required separating hyperplane. How do solve (33)? Once again we appeal to theorem 1 . So we need a separation algorithm for P^* . But this is easy. Given $y \in R^n$, $y \in P^*$ iff $\max\{yx | x \in P\} \leq 1$. Since our P is rational we can solve SOP. When we consider the general convex bodies, things are not so easy. Problem is that we cannot solve strong optimization problem. Recall that ellipsoidal algorithm, any localization algorithm for that matter, can not solve strong optimization problem by itself. This is the content of one of the main theorems of

Judin-Nemirovskii [J 7]. In the case of linear programming, we can solve strong optimization problem by utilizing underlying linear structure and rationality. For the general convex body, in order to obtain a separation algorithm for K^*, given $y \in R^n$, one solves

$$\max \{yx \mid x \in K\} , \tag{34}$$

with any desired accuracy. Let z be ε'-optimal solution of (34). After some manipulation one can show that $d(z,K) \leq \varepsilon$ if $yz \leq 1$, by choosing $\varepsilon' = \varepsilon r$, and using the fact that $u \in K^*$, $||u|| \leq 1/r$. Otherwise $zy \geq \varepsilon + zu$, $u \in K^*$. Let us note that only place, in our version i.e. using Theorem 2, we used r is to bound $||u|| \leq 1/r$, for $u \in K^*$. Consequently if one can bound the norm of the cuts used during the algorithm, then one can eliminate r completely. In Chapters 3,4,5 we used d, which substitutes 1/r in this sense.

There is a particular class of rational polytopes in combinatorial programming. Suppose K satisfying:

$$R^n_+ \cap S(0, r) \subseteq K \subseteq R^n_+ \cap S(0,R)$$

and (35)

$$x \in K, \quad 0 \leq y \leq x \Rightarrow y \in K$$

Antiblocker of K is defined by

$$A(K) = \{y \in R^n_+ \mid y^T x \leq 1, \quad \forall x \in K\}.$$

Following Grötchel-Lovász-Schrijver, we say a class is solvable if there is an polynomial algorithm for separation, hence optimization problem.

Theorem 6: Class of polytopes satisfying (35) is solvable. □

Now, we will study the applications of Theorem 5.

Matroids

Let E be a finite set, let $\mathcal{F} \leq \mathcal{P}(E)$, a subfamily of power set of E; which is closed under union and intersection. A function on $\mathcal{F}$ is submodular if

$$X, Y \in \mathcal{F} \Rightarrow f(X) + f(Y) \geq f(X \cap Y) + f(X \cup Y). \tag{36}$$

We will assume $f(\phi) = 0$, $X \subseteq Y \Rightarrow f(X) \leq f(Y)$, and f is integer valued. We associate a polyhedron to f

$$P(f) = \{x \in \mathcal{R}_+^E : x(S) \leq f(S), \ \forall S \in \mathcal{F}\}, \tag{37}$$

where $x(S) = \Sigma\{x_i : i \in S\}$.

When $\mathcal{F} = \mathcal{P}(E)$, if $f(S) \leq |S|$, then f is the rank function of a matroid (see Welsh [Z18]), otherwise f defines a polymatroid. It is known that vertices of P(f) are integer valued, when f defines a (poly)-matroid [Z16]. There is a very elegant algorithm, namely greedy algorithm to solve

$$\max \{cx | x \in P(f)\}. \tag{38}$$

When f defines a (poly)-matroid [Z16,Z18]. Thus by Theorem 5, we have a separation algorithm for P(f). Which in turn implies polynomial solvability of

$$\max \{cx | x \in P(f_i), \ i = 1,\ldots,k\}, \tag{39}$$

for a set of (poly)-matroid defining submodular functions. When $k = 2$, $P(f_1) \cap P(f_2)$ has integral vertices [Z16], but when $k \geq 3$, vertices are no longer integral. However Theorem 5 and Theorem 1 provides a polynomial algorithm for solution of (39), where the solutions are in general rational.

As mentioned above Theorem 5 provides a polynomial separation algorithm for matroid polyhedra. Recently, Cunningham [C39] developed a combinatorial algorithm for this purpose. It employs flow algorithms and matroid intersection theorem.

Submodular functions play a unifying role in combinatorics. Lovász [L 15] gave a unified review of submodularity, and its use in every branch of mathematics and discussed its relationship with convexity. Submodular flow problems are introduced by Edmonds and Giles [Z17]. Recently Lawler and Martel [L4-6], and Lawler introduced some generalization of classical flow problems: replacing capacity constraints by submodular ones.

So, given $\mathcal{F} \subset \mathcal{P}(E)$, closed under union and intersection, $f: \mathcal{F} \to \mathbb{Z}$, submodular, we would like to solve

$$\min \{f(X) \mid X \in \mathcal{F}\}. \tag{40}$$

We will assume that we know an upper bound on $|f(X)|$, $X \in \mathcal{F}$, say B and we have subroutines (which work in polynomial time) to decide whether $X \in \mathcal{F}$, with $x_1 \in X$, $x_2 \notin X$ w.l.o.g. assume $\cap\{X: X \in \mathcal{F}\} = \phi$, $\cup\{X: X \in \mathcal{F}\} = E$. Since we have an upper bound on $|f(X)|$, and f is integral we have to perform $\log_2 2B + 1$ binary search; i.e. we have to solve the decision problem

$$\exists\, X \in \mathcal{F} \quad \text{s.t.} \quad f(X) < K . \tag{41}$$

By translating $f(X) \leftarrow f(X) - K$, we reduce to problem of deciding whether $\exists\, X \in \mathcal{F}$ s.t. $f(X) < 0$. w.l.o.g. we can assume $f(\phi) = 0$. Consider the transformation

$$g(X) = f(X) + 2B|X| \;, \quad X \in \mathcal{F} .$$

This makes g a non-negative, monotone, submodular integral function. A function with above properties defined on $\mathcal{P}(E)$ is a polymatroid function. Now we will extend g to $\mathcal{P}(E)$; for $X \subset E$, define

$$\bar{X} = \cap \{X' \in \mathcal{F} \mid X \subseteq X'\} .$$

This is well-defined. Since $\mathcal{F}$ is closed under intersection, moreover by our assumptions this can be achieved in polynomial time. Let

$h(X) = g(\bar{X})$.

Clearly for $X \in \mathcal{F}$, $h(X) = g(X)$. h is a polymatroid function. Now $f(X) < 0 \iff g(X) < 2B|X|$, for $X \in \mathcal{F}$. If $h(X) < 2B|X|$, $X \in \mathcal{P}(E)$, then there is $X' \in \mathcal{F}$ such that $h(X) = g(X')$. Thus deciding whether $f(X) < 0$, $X \in \mathcal{F}$ is the same thing as whether $h(X) < 2B|X|$, for some $X \in \mathcal{P}(E)$. Let $\bar{x} = 2B\bar{1}$, where $\bar{1} \in \mathcal{R}^E$, with all components equal to 1. Then $h(X) < 2B|X|$, for some $X \in \mathcal{P}(E) \iff \bar{x} \notin P(h)$, since $\bar{x} \in P(h) \Rightarrow \bar{x}(S) = 2B|S| \le h(S)$. But clearly the latter problem is solvable since greedy algorithm solves the optimization problem.

Edmonds-Giles submodular flow problem is

$$\begin{aligned} &\max \Sigma\{c(e)x(e) : e \in E\} \\ &\text{s.t.} \\ &\text{i)}\ \ d(e) \le x(e) \le b(e) \\ &\text{ii)}\ \Sigma\{x(e) : h(e) \in S\} - \Sigma\{x(e) : t(e) \in S\} \le f(S), \quad S \in \mathcal{F} \end{aligned} \tag{42}$$

for some directed graph $G = (V,E)$, where $\mathcal{F} \subseteq \mathcal{P}(V)$ satisfying certain properties. It suffices to exhibit a polynomial separation algorithm in order to show that this problem is solvable. i) is trivial. For ii) one uses the previous algorithm. Recently Cunningham and Frank [C 38] gave a (polynomial) combinatorial algorithm for this problem.

Independence sets in perfect graphs

Class of perfect graphs is a nice example exhibiting difficulty of combinatorial problems. We know complete description of the facial structure of the underlying polytope for the independence number. But straightforward application of the ellipsoid or simplex method is out of question: i) number of constraints are exponentially large, ii) we do not have a polynomial separation algorithm.

Let $G = (V,E)$ be undirected graph, and $\bar{G} = (V,\bar{E})$ be its complement. A set $S \subseteq V$ is independent in G if there is no edge with both ends in S. $\alpha(G)$ is the size of maximum independent set of G. A set $C \subset V$ is a <u>clique</u> if every pair of nodes in C are adjacent in G, i.e. G[C], the graph induced by C is a complete graph. Clearly S is independent in G $\Longleftrightarrow$ S is a clique in $\bar{G}$. A k-clique cover is a partition of V into k cliques of G. Minimum such k is called <u>clique cover number</u> $\rho(G)$. Size of the maximum clique is called <u>clique number,</u> $\omega(G)$. Similarly a coluring of G is a partition of V into independent subsets of G. Size of the minimal such partition is called chromatic number, $\chi(G)$. A graph is perfect $\Longleftrightarrow$

$$\alpha(G[X]) = \rho(G[X]) \Longleftrightarrow \omega(G[X]) = \chi(G[X]), \quad \forall X \subseteq V . \tag{43}$$

The calculation of $\alpha(G)$ can be cast as integer program

$$\begin{aligned} &\max \ \Sigma x_i = \alpha(G) \\ &\Sigma\{x_i : i \in C\} \le 1 , \quad \forall C \in \mathcal{C} \\ &x \in \{0,1\}^V , \end{aligned} \tag{44}$$

where $\mathcal{C}$ is the set of all cliques. $\alpha^*(G)$ is the optimal value of (44) when integer restrictions are removed. The following is well-known [C14,L12]

<u>Theorem 7</u>: For a perfect graph G,

$$\alpha(G[X]) = \alpha^*(G[X]), \quad \forall X \subseteq V .$$

If we could solve (44) as an LP in polynomial time then we are done. However, since we do not have a separation algorithm, we will persue another line.

Lovász in 1979 [L14] defined a quantity for any graph. Let us call it <u>Lovász number</u>, $\vartheta(G)$. His purpose was to estimate the Shonnon capacity of a graph. Lovász showed that

$$\alpha(G) \le \vartheta(G) \le \alpha^*(G). \tag{45}$$

For perfect graphs inequalities become equalities, thus it suffices, to compute the Lovász number. Let

$$\mathcal{B} = \{B \in \mathcal{R}^{n\times n} \;\; trB = 1,\; B = B^t,\; B \text{ positive definite,}$$
$$b_{ij} = 0 \text{ if } (i,j) \in E\}\;, \tag{46}$$

where tr means trace, and for simplicity let $V = \{1,2,\ldots,n\}$. Then

$$\vartheta(G) = \max\{\Sigma b_{ij} \mid B \in \mathcal{B}\}. \tag{47}$$

One can show that $\mathcal{B}$ is a convex set, hence (47) is a convex programming problem. In order to apply ellipsoidal algorithm, say via Theorem 1, one needs a separation algorithm. Grötchel-Lovász-Schrijver gave a separating algorithm for $\mathcal{B}$ using the Sylvester criteria of positive definiteness. Since $\alpha(G)$ is an integer for a perfect graph, one chooses $\varepsilon = 1/2$. If one wants to solve weighted version of the above problem, that will change only the objective function to

$$\sum_{i,j} \sqrt{\omega_i \omega_j}\; b_{ij}\;,$$

of course one should replace square root term with suitable rational approximation.

Since G is perfect $\Longleftrightarrow$ $\bar{G}$ is perfect, and $\alpha(G) = \omega(\bar{G})$, the computation of $\alpha(G)$, $\omega(G)$, $\rho(G)$, $\chi(G)$ for perfect graphs is polynomial.

Since the above algorithm solves the optimization problem of $\alpha(G)$ for perfect graphs, consequently it also solves the separation problem; one can appeal to Theorem 5 or 6. There is a more direct method. Given $y \geq 0$ solve the maximum weighted clique problem for G with weights given by y. If the optimal weight is ≤ 1, then y is in the polytope defined by (44). Otherwise optimal solution will give a cut. In order to solve the maximum weighted clique problem for G, one can solves maximum stable (independent) set problem for $\bar{G}$, since $\bar{G}$ is also perfect.

NP-Completeness

A natural question to ask is the following: what is the importance of the ellipsoidal algorithm in settling the question *P* = *NP*, if any. Well, if you can solve any *NP*-complete problem by ellipsoidal algorithm, then you have a proof that *P* = *NP*. In order to solve any optimization problem by ellipsoidal algorithm you need a polynomial separation algorithm. Similarly, if you have a polynomial optimization algorithm, then you have a polynomial separation algorithm. For combinatorial problems using LP approach, this amounts to being able to recognize facet defining inequalities in polynomial time. Above remarks lead to following theorem, which has been observed by several researchers [G48,K9 ,P5].

Theorem 8 : For an *NP*-complete problem, there is no polynomial separation algorithm unless *P* = *NP*.

Karp and Papadimitrou [K9] made another important observation. Recall that the decision problem C $\in$ *NP* if given an instance of C and some extra information, say a candidate solution, we can verify the solution with polynomial effort if indeed the candidate is a solution. For example, given a graph G = (V,E) and a subset of edges E' we can easily check whether edges in E' form a Hamiltonian circuit. A decision problem C $\in$ co-*NP* if we can verify with polynomial effort, if the answer to our decision problem is no. In the above example, Hamiltonian circuit problem is in Co-*NP*, if for given a graph G, we can verify with polynomial work that G does not contain an Hamiltonian circuit. If one can show that for any *NP*-complete problem, the complementary question can be verified with polynomial effort, then it would follow that *NP* = co-*NP* . Karp and Papadimitrou proved the following,

Theorem 9 : If any *NP*-complete problem has valid inequalities whose coefficients are well-bounded and whose verification can be done with polynomial effort then *NP* = co-*NP* .

□

8 Yamnitsky–Levin algorithm

In 1982 B. Yamnitsky and L.A. Levin [Y 3] announced another polynomial algorithm for linear programming. It is a modification of A.Ju.Levin's 1965 [L 11] method of center of gravity (MCG) (see Chapter 3). MCG works with polytopes. It checks centroid of the current polytope. If it is not a solution then by a suitable hyperplane, algorithm discards part of the polytope, and takes remaining polytope as the new one, and thus continues. Since polytopes become messy through bisection by a hyperplane passing through their center of gravity, A.Ju.Levin suggested that in every q steps, for some q, one could replace the polytope with a simplex of higher volume. It is known that there exist constants γ_n depending on dimension n such that the volume of the simplex enclosing the polytope is $\leq \gamma_n$ times the volume of the polytope. Yamnitsky, in this thesis, [Y 2] showed that q can be taken as 1 for any n, moreover the volume of the simplex enclosing the desired part of the polytope will decrease by a ratio $\leq \exp\{\frac{-1}{2(n+1)^2}\}$ provided that the original polytope was a simplex. Hence the new algorithm works with simplices. It checks the center of the current simplex. If it is not a solution, by taking a suitable hyperplane divides the simplex into two parts, and forms a new simplex containing the desired part. Naturally, the new simplex has a smaller volume. It has some similarities with the ellipsoidal algorithm. Its complexity is roughly n times that of ellipsoidal algorithm. It is very difficult to find the new simplex with minimal volume. But it does not require square roots, so it is relatively easy to implement in exact arithmetic. It seems more compatible with simplex method.

8.1 PRELIMINARIES

The points $x^1, x^2, \ldots, x^{k+1} \in R^k$ are affinely independent if

$$\sum_{i=1}^{k+1} \alpha_i x^i = 0 \ , \qquad \sum_{i=1}^{k+1} \alpha_i = 0 \tag{1}$$

implies that $\alpha_i = 0$, $i = 1,\ldots,k+1$. Equivalently $\{x^1,\ldots,x^{k+1}\}$ are affinely independent if and only if $\bar{x}^i = \binom{x^i}{1} \in R^{k+1}$, $i = 1,\ldots,k+1$ are linearly independent. This is so iff

$$\det \begin{bmatrix} x^1 & x^2 & & x^{k+1} \\ 1 & 1 & & 1 \end{bmatrix} \neq 0. \tag{2}$$

In this case,

$$co(x^1,\ldots,x^{k+1}) = \{x: x = \sum_{i=1}^{k+1} \alpha_i x^i \ , \ \Sigma\alpha_i = 1, \ \alpha \geq 0\} \tag{3}$$

is a k-simplex whose extreme points are precisely $x^1, x^2, \ldots, x^{k+1}$. Thus a k-simplex in R^k defines a $(k+1)\times(k+1)$ nonsingular matrix. It is well-known that affinely independent $x^1, x^2, \ldots, x^k$ in R^k define a hyperplane, say, $\Pi x = \Pi^o$ such that $\Pi x^i = \Pi^o$, $i \leq k$, and $\Pi \in R^k$, $\Pi^o \in R$ are uniquely determined. Hence for each k-element subset of $x^1, x^2, \ldots, x^{k+1}$, we have a hyperplane of the form $\Pi x = \Pi^o$. Hence the vectors $(\Pi, \Pi^o) \in R^{k+1}$, one for each k-element subset of $x^1, x^2, \ldots, x^{k+1}$ will define a $(k+1)\times(k+1)$ nonsingular matrix. Conversely any $(k+1)\times(k+1)$ nonsingular matrix T defines a cone in R^{k+1} by the relation

$$T(x) \geq 0 \tag{4}$$

with apex at $\underline{0}$, i.e. cone is $\{x: T(x) \geq 0\}$. By intersecting this cone with an hyperplane in "general position", $ex = 1$ we obtain a k-simplex. By general position we require the following. $\forall i \leq k+1$, the system,

$$\begin{aligned} &T_j x = 0, \quad j \neq i \quad , \quad j \leq k+1 \\ &ex = 1 \end{aligned} \tag{5}$$

has a unique solution x^i, where T_j denotes the j'th row of T. Note also that $co\{x^1, x^2, \ldots, x^{k+1}\}$ is a (k+1)-simplex in R^{k+1}, and $co\{x^1, x^2, \ldots, x^{k+1}\}$ is a face of this simplex, which itself is a k-simplex.

Thus the equation

$$T_i x = 0 \tag{6}$$

is satisfied by all x^j, $j \neq i$, and is called the face opposite to x^i.

We define the standard simplex S_n as

$$S_n = \{x \in R^n : \sum_{i=1}^{n} x_i \leq 1, \quad x \geq 0\} . \tag{7}$$

Clearly we have

$$S_n = co\{e_1, e_2, \ldots, e_n, 0\} \tag{8}$$

where e_i's are the usual unit vectors. To every $x \in S_n$ we have $\alpha \in R^{k+1}$ satisfying

$$\begin{aligned} &\alpha \geq 0 , \\ &\sum_{i=1}^{n+1} \alpha_i = 1 , \\ &x = \sum_{i=1}^{n} \alpha_i e_i . \end{aligned} \tag{9}$$

Clearly the mapping $(x,0) \longleftrightarrow \alpha$ is 1-1 with determinant 1. One can easily show that by elementary methods

$$\mathrm{vol}(S_n) = \int_{S_n} dx = \frac{1}{n!} \quad .$$

For any n-simplex $S = \mathrm{co}\{x^1, x^2, \ldots, x^{n+1}\}$ the mapping

$$\alpha \overset{T}{\longleftrightarrow} \sum_{i=0}^{n} \alpha_i \binom{x_i}{1}$$

will provide a 1-1 correspondence between S_n and S. Thus, we have

$$\mathrm{vol}(S) = \mathrm{vol}(S_n) \det T = \frac{\det T}{n!} \qquad (10)$$

The Problem

Let us consider the problem

Find x such that

$$\text{LI} \quad Ax \geq b \qquad (11)$$

where $A \in Z^{m \times n}$, $b \in Z^m$, $m \geq n$. Let

$$P = \{x \in R^n : Ax \geq b\}. \qquad (12)$$

Clearly, by introducing a new variable t, we can write (11) as

$$[A,-b]\binom{x}{t} \geq 0$$

$$t > 0 \ . \qquad (13)$$

Let

$$\tilde{P} = \{(x,t) : [A,-b]\binom{x}{t} \geq 0,\ t > 0\} \ . \qquad (14)$$

One has an equivalence between P and $\tilde{P}$ in the sense that $\forall (x,t) \in \tilde{P} \Rightarrow (\frac{x}{t},1) \in \tilde{P} \Rightarrow \frac{x}{t} \in P$, and vice versa $x \in P \Rightarrow (x,1) \in \tilde{P}$.

Let $\bar{P} = \{(x,1): x \in P\}$. We will work with $\bar{P}$ instead of P. The advantage of $\bar{P}$ is that the equations become homogenous. This process is known as homogenization see, e.g. [B2], [S40].

By making the following substitution

$$A \longleftarrow [A,-b] \ , \ x \longleftarrow (x,t) \tag{15}$$

we will represent system LI as

$$\begin{aligned} &Ax \geq 0 \\ &x_{n+1} = 1. \end{aligned} \tag{16}$$

8.2 THE ALGORITHM

We will start with a simplex S containing at least some solutions of

$$\begin{aligned} &Ax \geq 0 \\ &x_{n+1} = 1. \end{aligned} \tag{1}$$

We will discuss the choice of the initial simplex later. Thus we have $x^1, x^2, \ldots, x^{n+1}$ being the vertices of S, and $T_i x = 0$, $i = 1,\ldots,n+1$, being the face in fact facet opposite to x^i. The <u>center</u> or centroid of the simplex $\bar{x} = \bar{x}(S)$ is defined as

$$\bar{x} = \frac{1}{n+1} \sum_{i=1}^{n+1} x^i. \tag{2}$$

If $\bar{x}$ is a solution, we are done. Otherwise let $a \in R^{n+1}$ be a row of A for which

$$a\bar{x}_s < 0. \tag{3}$$

Choose i_o such that

$$ax^{i_o} = \max_i \{ax^i\} . \qquad (4)$$

Update T_{i_o} as

$$T_{i_o} \leftarrow T_{i_o} + \frac{T_{i_o} x^{i_o}}{ax^{i_o}} \, t \, a . \qquad (5)$$

Since new T_i's will define a new simplex, we can continue the process.

Thus we can summarize the algorithm as

0. Initialize. Choose a simplex s containing solutions of the system, if any.
1. Find the center of the simplex. If the center is a solution, stop. Otherwise choose a violated constraint. Choose a vertex of the simplex by (4).
2. Update T_{i_o} by (5) for a suitable t.
3. Go to 1.

Let us recall a fundamental result in linear programming, i.e. Farkas Lemma

Lemma 1: For $y \in R^m$, $y \geq 0$, $yA \geq \alpha$ is a valid inequality for $Ax \geq b$ if $\alpha \leq yb$. □

Proposition 2: Let $dx \geq 0$ be a valid inequality for $\bar{P}$, and $ax \geq 0$ be another valid inequality, in particular a constraint of $\bar{P}$, then $\forall t > 0$, $(d + ta)x \geq 0$ is a valid inequality.

Proof: By assumption, for $x \in P$ we have $dx \geq 0$, $ax \geq 0$, then $(d + ta)x = dx + t\,ax \geq 0$. □

Determination of vertices of the new simplex

For notational convenience, let $d = T_{i_o}$, where i_o is defined by (4). Since $dx = 0$ is the opposite face to x^{i_o}, x^j, $j \neq i_o$ lies on the hyperplane defined by $dx = 0$, and x^{i_o} does not lie on it. Thus, for the new simplex S', $x'^{i_o} = x^{i_o}$. For $j \neq i_o$ vertex x'^j will lie on the line

$$x^{i_o} + \tau_j(x^j - x^{i_o}),$$

τ_j is chosen so that $d'x'^j = 0$, where

$$d' = d + \frac{dx^{i_o}}{ax^{i_o}} t\ a . \tag{6}$$

Since

$$d'x'^j = (d + \frac{dx^{i_o}}{ax^{i_o}} t\ a)(x^{i_o} + \tau_j(x^j - x^{i_o}))$$

$$= dx^{i_o} + \tau_j\ d(x^j - x^{i_o}) + t\ \frac{dx^{i_o}}{ax^{i_o}}\ a\ x^{i_o} +$$

$$+ t\ \tau_j\ \frac{dx^{i_o}}{ax^{i_o}}\ (a\ x^j - a\ x^{i_o}) .$$

Set

$$\varepsilon_j = ax^j/ax^{i_o}\ ,\ j \neq i_o . \tag{7}$$

Then

$$d'x'^j = dx^{i_o}\{(1 - \tau_j) + t + t\ \tau_j(\varepsilon_j - 1)\}.$$

Thus

$$d'x'^j = 0 \iff$$

$$1 + t = [1 + (1 - \varepsilon_j)t]\tau_j . \tag{8}$$

Hence

$$\tau_j = \frac{1+t}{1+t-t\varepsilon_j} \quad , \quad j \neq i_o \tag{9}$$

Let us make the substitution

$$t \longleftarrow \frac{t}{1+t} \quad . \tag{10}$$

Note that this maps $(0,\infty)$ into $(0,1)$. Thus we have

$$\tau_j = \frac{1}{1-t\varepsilon_j} \quad , \quad j \neq i_o \; . \tag{11}$$

Thus vertices x'^j of the new simplex S' are

$$\begin{aligned} x'_{i_o} &= x_{i_o} \\ x'_j &= x_j + \tau_j(x_j - x_{i_o}) , \quad j \neq i. \end{aligned} \tag{12}$$

Let us introduce the index sets

$$\begin{aligned} J^- &= \{j: \varepsilon_j < 0\} \\ J^o &= \{j: \varepsilon_j = 0\} \\ J^+ &= \{j: \varepsilon_j > 0\} \; . \end{aligned} \tag{13}$$

Notice by assumption $a\bar{x} < 0$, thus $J^- \neq \phi$. If $J^o \cup J^+ \neq \phi$, there is no solution, since $ax \geq 0$ does not intersect S. If $J^+ = \phi$, $J^o \neq \phi$, then

$$P \subset \{x: a_i x = b_i \; , \; i \in J^o\} \; .$$

In this case one has to work in smaller dimension.

Thus let us assume $J^+ \neq \phi$. Then $i_o \in J^+$ and $\varepsilon_{i_o} = 1$, $\varepsilon_j \leq 1$, $j \leq J^+$. Define

$$\tilde{J} \equiv J^+ - \{i_o\} \tag{14}$$

and

$$\begin{aligned} \sigma_j &= \varepsilon_j , \quad j \in J^+ \\ \gamma_j &= -\varepsilon_j , \quad j \in J^- . \end{aligned} \tag{15}$$

Since $a\bar{x} < 0$ we have

$$\sum_{j\in J^-} \gamma_j > \sum_{j\in J^+} \sigma_j . \tag{16}$$

Defining

$$\gamma = \Sigma\{\gamma_j : j \in J^-\} , \quad \sigma = \Sigma\{\sigma_j : j \in \tilde{J}\} , \tag{17}$$

we have

$$\gamma > \sigma + 1 . \tag{18}$$

Let us compute vol(S'). By (1.10), we compute

$$\det\left[x_1', x_2', \ldots, x_n'\right] = \det\left[x_1 + \tau_1(x_1 - x_{i_o}), \ldots, x_{i_o}, \ldots \right.$$
$$\left. \ldots, x_n + \tau_n(x_n - x_o)\right]$$
$$= (\prod_{j\neq 0} \tau_j) \det [x^1, x^2, \ldots, x^n] .$$

Thus

$$\mathrm{vol}(S')/\mathrm{vol}(S) = \prod_{j \neq i_o} \tau_j . \tag{19}$$

In other words, as a function of t, the step size,

$$\mathrm{vol}(S')/\mathrm{vol}(S) = F(t),$$

where

$$F(t) = \Big(\prod_{j \in J^-} \frac{1}{1+t\gamma_j}\Big)\Big(\prod_{j \in \tilde{J}} \frac{1}{1-t\sigma_j}\Big). \tag{20}$$

When $\tilde{J} = \phi$, we immediately have

<u>Lemma 3</u>: If $\tilde{J} = \phi$, then t can be taken as arbitrarily large.

□

<u>Remark</u>: Letting $t = +\infty$ in (5) has the effect of changing

$$d'x = (d+ta)x \geq 0 \quad \text{to} \quad ax \geq 0.$$

Note due to transformation (10), our t will be in (0,1). Thus volume reduction will be

$$\prod_{j \in J^-} \frac{1}{1+\gamma_j} .$$

<u>Remark</u>: $\tau_j > 1$, for $j \in \tilde{J}$, $\tau_j < 1$, for $j \in J^-$. Moreover

$$\tau_j \leq \frac{1}{1-t}, \text{ since } \sigma_j \leq 1, \text{ for } j \in \tilde{J} .$$

Let

$$G(t) = \frac{1}{F(t)} = \prod_{j \in J^-} (1+t\gamma_j) \prod_{j \in \tilde{J}} (1-t\sigma_j) . \tag{21}$$

Since minimizing F(t) is the same thing as maximizing G(t), we might as well work with G(t). Note that it is not so-easy to find the maximizer of G(t) explicitly. However, we will maximize a related function.

Lemma 4: For $\sigma \in (0,1)$, $t \in (0,1)$

$$(1-t\sigma) \geq (1-t)^{\sigma}$$

Proof: Let $f(t) = (1-t\sigma)-(1-t)^{\sigma}$. Since $f(0) = 0$, $f(1) = 1-\sigma, >0$, it suffices to show $f'(t) \geq 0$, which is easy.

Lemma 5: For $t \geq 0$, $\gamma_j \geq 0$,

$$\prod_{j \in J^-} (1+t\gamma_j) \geq (1+t \sum_{j \in J^-} \gamma_j) . \qquad \square$$

For notational convenience, let $n \leftarrow n + 1$.
By repeated application of the above lemmas, we have

$$G(t) \geq (1 + (\gamma t)(1-t)^{\sigma} = H(t) \qquad (22)$$

where γ, σ are defined by (17). Let us recall that $\sigma_i \leq 1$, thus $\sigma \leq |J| \leq n-2$ and $\gamma > \sigma + 1$.

Let τ be defined so that $\sigma + \tau = n-2$. Then we have

Lemma 6: $(1+t\gamma)(1-t)^{\sigma} \geq (1+(\gamma + \tau)t)(1-t)^{\sigma+\tau}$

Proof: It is the same thing as proving

$$f(t) = (1+t\gamma) - (1+(\gamma + \tau)t)(1-t)^{\tau} \geq 0 ,$$

for $\tau > 0$. Since

$$\begin{aligned} f'(t) &= \gamma - (\gamma + \tau)(1-t)^{\tau} + \tau(1-t)^{\tau-1}(1+(\gamma + \tau)t) \\ &= \gamma - \gamma(1-t)^{\tau} + \tau(1-t)^{\tau} (\gamma + \tau)t \\ &= \gamma[1-(1-t)^{\tau}]+ \tau(1-t)^{\tau-1} [1+(\gamma + \tau)t-(1-t)] . \end{aligned}$$

Since

$$[1+(\gamma + \tau)t-(1-t)] = (\gamma + \tau + 1)t, \quad f'(t) > 0,$$

for $t\epsilon(0,1)$, $\tau \geq 0$. Thus

$$H(t) \geq (1+(\gamma + \tau)t)(1-t)^{\sigma+\tau} \geq (1+(n-1)t)(1-t)^{n-2} = H_1(t) \tag{23}$$

Maximization of $H_1(t)$ is a simple matter now. Since

$$H_1'(t) = (n-1)(1-t)^{n-2}-(n-2)(1+(n-1)t)(1-t)^{n-3},$$

$$H_1'(t) = 0 \iff (n-1)(1-t)-(n-2)-(n-1)(n-2)t = 0.$$

Thus

$$t = \bar{t} = \frac{1}{(n-1)^2} \,. \tag{24}$$

One can easily show that $H_1'(t) \leq 0$, for $t\epsilon[0,1]$.

Let us compute $H_1(t)$,

$$H_1(t) = (1+(n-1)t)(1-t)^{n-2} = (1 + \frac{1}{n-1})(1- \frac{1}{(n-1)^2})^{n-2} \,. \tag{25}$$

By multiplying and dividing by $(1 - \frac{1}{n-1})$, we have

$$H_1((\frac{1}{n-1})^2) = \frac{(1-(\frac{1}{n-1})^2)^{n-1}}{1- \frac{1}{n-1}} \,. \tag{26}$$

Recalling that

$$\log(1-x) = - \sum_{k=1}^{\infty} \frac{x^k}{k} \,, \text{ for } 0 < x < 1,$$

we have

$$\lambda = \log H_1\left(\left(\frac{1}{n-1}\right)^2\right) = \sum_{k=1}^{\infty} \frac{1}{k}\left(\frac{1}{n-1}\right)^k - (n-1)\sum_{k=1}^{\infty} \frac{1}{k}\left(\frac{1}{n-1}\right)^{2k}$$

$$= \frac{1}{n-1} - (n-1)\left(\frac{1}{n-1}\right)^2 + \frac{1}{2}\left(\frac{1}{n-1}\right)^2$$

$$- \frac{1}{2}\left(\frac{1}{n-1}\right)^3 + \sum_{k=3}^{\infty} \frac{1}{k} \underbrace{\left\{\left(\frac{1}{n-1}\right)^k - \left(\frac{1}{n-1}\right)^{2k+1}\right\}}_{>0}$$

$$> \frac{1}{2}\left(\frac{1}{n-1}\right)^2 - \frac{1}{2}\left(\frac{1}{n-1}\right)^3 > \frac{1}{2n^2} \quad . \qquad \square$$

Thus we have proved

Theorem 7: One can always achieve a volume reduction less than $\exp\left\{-\frac{1}{2(n+1)^2}\right\}$, by choosing $t = \frac{1}{n(n-2)}$ in (6). $\square$

Remark 1: n in the above formula is the original dimension.

Remark 2: t in (24) is the transformed t. Applying inverse mapping of (10) one obtain $\bar{t} = 1/n(n-2)$.

One can directly verify the volume reduction ratio of $\exp\left\{-\frac{1}{2n^2}\right\}$ for choice of $t = \bar{t} = \left(\frac{1}{n-1}\right)^2$. Let $\lambda_j = \frac{1}{\tau_j}$. Then

$$G(t) = \prod_{j \neq i_0} \lambda_j \ . \tag{27}$$

Since by (11), $\lambda_j = 1 - \bar{t}\varepsilon_j$, $\varepsilon_j \leq 1$, implies that

$$\lambda_j \geq 1 - \bar{t} \ . \tag{28}$$

On the other hand,

$$\Sigma\lambda_j = \sum_{j\neq i_o} (1-\bar{t}\varepsilon_j) = (n-1)-\bar{t}\,\Sigma\,\varepsilon_j = n-1-(\gamma - \sigma)\bar{t} > n-1-\bar{t}\ . \tag{29}$$

Let us consider the optimization problem

$$\begin{aligned} \min \quad & \Pi\lambda_j = f(\alpha) \\ \text{s.t.} \quad & \lambda_j \geq 1-\bar{t} = \mu \\ & \Sigma\lambda_j > n-1-\bar{t} = \alpha . \end{aligned} \tag{30}$$

Since $f(\alpha)$ is monotonic decreasing, we can replace the inequality sign with equality. Then Kuhn-Tucker points of the above optimization are permutations of the

$$(\underbrace{\mu,\mu,\ldots,\mu}_{n-1},\ \alpha-(n-2)\mu)\ ,$$

and

$$(\bar{\alpha},\ \bar{\alpha},\ldots,\ \bar{\alpha}),$$

where $\bar{\alpha} = \alpha/n$. Clearly latter stationary point has a larger objective function value, hence can be discarded. Since

$$\alpha - (n-2)\mu = n-1-\bar{t}-(n-2)(1-\bar{t}) = 1+(n-1)\bar{t},$$

and

$$f(\alpha) \geq (1+(n-1)\bar{t})(1-\bar{t})^{n-2} = H_1(\bar{t}),$$

we are back to equation (26).

Since

$$H(t) = (1+\gamma t)(1-t)^{\sigma},$$

a good choice for t can be found by equation $H'(t) = 0$, which gives

$$\tilde{t} = \frac{\gamma - \sigma}{\gamma(\sigma + 1)} \quad . \tag{31}$$

Notice that $H(t)$ is concave. This most easily can be seen by looking at $(\log H(t))''$. Thus

$$H(t) \le H(\tilde{t})$$

in particular

$$H(\bar{t}) \le H(t). \tag{32}$$

Since

$$\exp \{\frac{1}{2n^2}\} \le H_1(\bar{t}) \le H(\bar{t}) \le H(\tilde{t}), \tag{33}$$

the choice of $t = \tilde{t}$ by (31) will guarantee a volume reduction of $\exp \{- 1/2(n+1)^2\}$, where again n is the original dimension.

Note that we made the transformation given by (10). Applying corresponding inverse transformation we have

$$\bar{t} = \frac{1}{n(n-2)}$$

$$\tilde{t} = \frac{\gamma - \sigma}{\sigma(\gamma+1)} \quad .$$

We have included two figures to show how the algorithm works geometrically. Figure 8.1 illustrates a typical iteration in the plane, where as Figure 8.2 illustrates the case $\tilde{J} = \phi$.

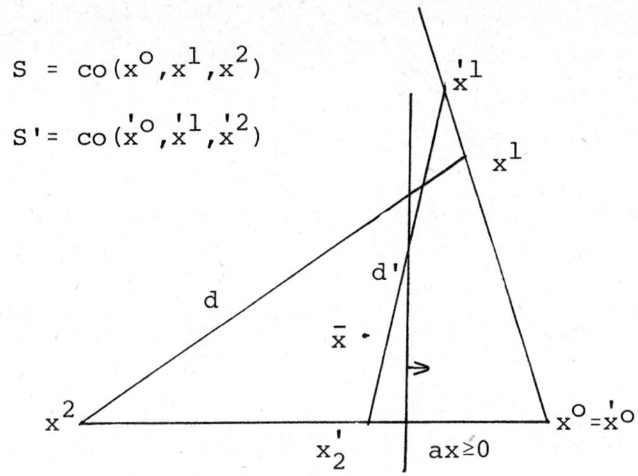
S = co(x^o,x^1,x^2)
S'= co(x'^o,x'^1,x'^2)
x'^1
x^1
d'
d
x̄
x^2
x'_2
ax≥0
x^o=x'^o

Fig. 8.1

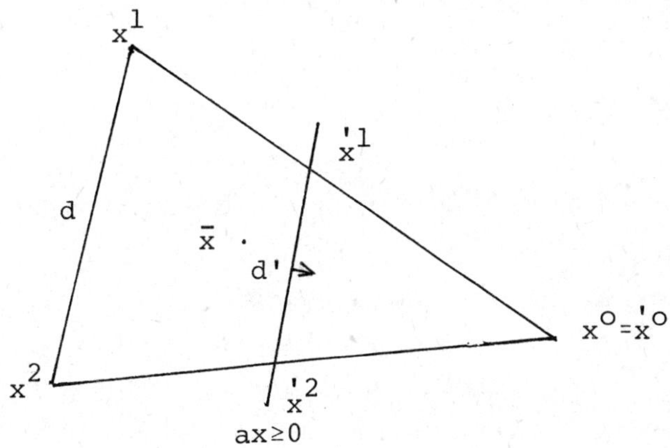
x^1
x'^1
d
x̄
d'
x^o=x'^o
x^2
x'^2
ax≥0

Fig. 8.2

8.3 INITIAL SIMPLEX AND CONVERGENCE

Since the basic idea of the algorithm, like ellipsoidal algorithm, is to reduce the volume of the region, which con contains all or some of the solutions it is desirable to start the algorithm with an initial simplex having smallest possible volume with reasonable effort. For most of the LP problems nonnegativity constraints are very natural. One can add one more hyperplane to form a simplex. If one has an equation of the form

$$\sum_{i=1}^{n} \alpha_i x_i \le \beta , \tag{1}$$

where $\alpha_i > 0$, $\beta > 0$, as a constraint or as a valid inequality derived from the constraints using Farkas' lemma, one can use it. In the worst case, one can take the initial simplex as

$$x_i \ge 0 , \quad i = 1,\dots,n \tag{2}$$

$$\sum^{n} x_i \le R$$

$$x_{n+1} = 1,$$

where R is large enough. An a priori R, when one considers the system as a black box, would be $R = 2^L$ as in the ellipsoidal algorithm.

One can choose a better initial simplex, at least in terms of its volume, if one is willing to do some extra work. Most often, a subset of constraints will define a simplex. This simplex automatically will contain all the solutions since it is a relaxation of the original system. By using Farkas' lemma one can find n+1 valid inequality in "general position" which will define a simplex. If one uses homogenization, then one must make sure that the equation ex = 1, is in general position as explained in Section 1. In this case any n+1, linearly independent rows of [A,-b] , or B[A,-b], where B is non-negative will define a cone in R^{n+1}, whose intersection with a hyperplane in general position will define an n-simplex. Yamnitsky-Levin work through the matrix B. They assert that any non-negative B satisfying det BA ≥ 0, BA $x \geq 0$ will define a simplex, if one chooses $e = (1,1,\ldots,1)BA$. They implicitly assume $e > 0$. Of course, this will guarantee the above requirement. In general, an arbitrary choice of e, in the form $e^t = f^tA$, $f > 0$, $f \in R^n$, will not guarantee that e is in general position.

<u>Ex</u>: Consider strict linear inequality system

$$x_1 > 1$$

$$x_1 < 2$$

$$x_2 > 1$$

$$x_2 < 2$$

$$t > 0$$

One can represent is as $Ax \geq 0$, where, $x = (x_1,x_2,t)$,

$$A = \begin{bmatrix} 1 & 0 & -1 \\ -1 & 0 & 2 \\ 0 & 1 & -1 \\ 0 & -1 & 2 \\ 0 & 0 & 1 \end{bmatrix}$$

clearly for

$$B = \begin{bmatrix} 1 & 0 & 0 & 0 & 0 \\ 0 & 1 & 0 & 0 & 0 \\ 0 & 0 & 1 & 0 & 0 \end{bmatrix}$$

$$\det BA \neq 0, \quad e = (1,1,1,1,1)\,A = (\,0,0,3)$$

$$\begin{array}{ll} BAx \geq 0 & x_1 \geq 1/3 \\ & x_1 \leq 2/3 \\ ex = 1 & x_2 \geq 1/3 \\ & t = 1/3 \end{array}$$

which will not define a simplex. But, if one chooses e as e = (1,1,1)BA, then the resulting system is

$$\begin{array}{ll} x_1 - t \geq 0 & (1) \\ -x_1 + 2t \geq 0 & (2) \\ x_2 - t \geq 0 & (3) \\ x_2 = 1 & \end{array}$$

will define a simplex, shown in Fig. 8.3

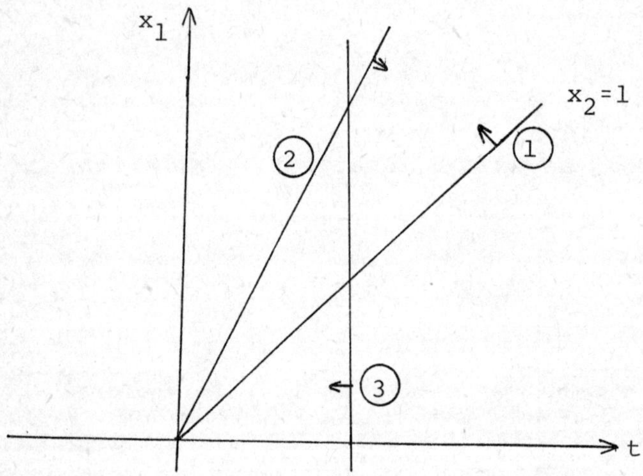

Fig. 8.3

Convergence of the algorithm

One can easily evaluate the volume of the initial simplex. An a priori upper bound is, in the notation of ellipsoidal algorithm

$$\frac{1}{n!}\left(\frac{2^L}{n}\right)^{(n+1)} = \frac{2^{L(n+1)}}{n!\,n^{n+1}} \; . \qquad (3)$$

If one assumes full-dimensionality, then set of solutions, provided $P \neq \phi$, is by (1.10), see, e.g. [G13]

$$\frac{1}{n!}\, n^{n+1}\, 2^{-L(n+1)} \; . \qquad (4)$$

Since the volume reduction is at least

$$e^{\frac{-1}{2(n+1)^2}} < 2^{\frac{-1}{2(n+1)^2}} \qquad (5)$$

in,

$$\left(\log \frac{vol(S^o)}{vol(P)}\right) 2(n+1)^2 = 2(n+1)^2\,[2(n+1)L + 2(n+1)\log_2 n] \qquad (6)$$

$$= 0(n+1)^3 L), \qquad (7)$$

iterations either one finds a solution to strict inequality system, or verifies that there is no solution.

For inequality-equality system, even though Gács-Lovász' perturbation technique remains available, one does not need it. Let us recall that A.Ju.Levin [L11] devised his method of center of gravity for convex programming. Similarly, this new algorithm is valid for unconstrained and constrained convex programming provided the cuts are taken as described in Chapter 3 and Chapter 5. Proofs of Hachiyan, and Judin-Nemirovskii carry over automatically.

In order to be complete, let us described the algorithm for unconstrained convex minimization problem. Given $\min_{x \in G} f(x)$, find a simplex containing G, or part of G, for which at least one minimizer is known to be contained in. At the center of simplex $\bar{x}(S)$ check whether $0 \in \partial f(\bar{x}(S))$. If that is the case stop, an optimal solution is found. Otherwise take $a \in \partial f(\bar{x})$, apply the cut $ax \leq a\bar{x}$ as the violated constrained. For the given accuracy $\varepsilon > 0$, continue the algorithm untill volume of the current simplex is less than or equal to $\omega_n \varepsilon^n/d^n$, where d is the Lipschitz constant of f, and ω_n is the volume of the unit ball in R^n. Notice that this estimates are given before the homogenization is described. Obviously, it is computationally convenient to work in R^{n+1} through homogenization.

8.4 RELATIONSHIP WITH ELLIPSOIDAL ALGORITHM

As it is apparent from the description both fall in the same category of algorithms: they both try to localize the solutions. They reduce the volume of localization each working within a class of convex bodies. Volume reduction is only dependent, in the worst case, to the dimension of the problem. That is problem characteristic can not worsen the given upper bounds. In terms of worst case behaviour, the new algorithm is n times slower than ellipsoidal algorithm.

One advantage of Yamnitsky-Levin algorithm is that it does not require square roots. Thus, it allows exact arithmetic through some elaborate book-keeping methods. Deep cuts are almost automatic in the new algorithm. If one just take the step size in (2.5) as $1/n(n-2)$, (while working in R^{n+1}) as suggested by Yamnitsky-Levin, that roughly corresponds to original step size of Hachiyan and Judin-Nemirovskii.

Parallel cuts are possible in ellipsoidal algorithm even though one has to do some arithmetic. In the new algorithm it is not possible.

Surrogate cuts are possible in the new algorithm as in the ellipsoidal algorithm.

When applied to a convex minimization problems both algorithms can be considered within the class of subgradient algorithms. Thus they do not decrease the objective function, since try to minimize the volume of location of optimal solutions. Yamnitsky-Levin claim that their algorithm is "expected" to be a descent algorithm. Obviously, there is no evidence to support it, and naturally it is not true.They make some remarks that their algorithm is better suited, in the sense that it is relatively easy to check whether a simplex contains all the solutions, etc. We find no evidence to support that claim. Even though it seems a simplex is more manageable than an ellipsoid, in R^n both are hard to work. It seems one should be able to work with simplices more easily because of large literature on polyhedral and linear programming theory.

How does one solve linear programming problems with the new algorithm? Obviously all the methods in Chapter 4 automatically carry over. At this point we do not know whether we could substitute the ellipsoidal transformation of objective function, with T c, where T defines the simplex. Recall that each ellipsoid, either defines a new metric, or gives you a transformation between two copies of R^n . Our limited experiments with the ellipsoidal method gives the impression that ellipsoid obtained after sufficiently large iterations capture the "essence" of the underlying polytope. There is some theoretical results in this vein see Goffin [G 37-38] . Whether such a statement holds for the new algorithm remains to be seen.

References

The following abbreviations are used throughout references:

EMM Ekonomika i Matematicheskie Metody
MP Mathematical Programming
SMD Soviet Mathematics Doklady
CMMP USSR Computational Mathematics and Mathematical Physics
(M)OR (Mathematics of) Operations Research

[A1] A.I. Aćil'diev, The method of normal solutions in linear programming, in S.H. Lavington ed. Information Processing 80 (North-Holland, Amsterdam, 1980), 667-670.

[A2] I. Adler, R.P. Mclean, and J.S. Provan, An application of the Khachian-Shor algorithm to a class of linear complementary problems, Cowles Foundation discussion paper no.549, 1980.

[A3] S. Agmon, The relaxation method for linear inequalities, Canadian Journal of Mathematics 6 (1954), 382-392.

[A4] A.V. Aho, J.E. Hopcroft, and J.D. Ullman, The design and Analysis of Computer Algorithms, (Addison-Wesley, Reading, Mass. 1976).

[A5] M. Akgül, On Shor-Hachian algorithm, University of Waterloo Research Report CORR 80-50.

[A6] ———— , An ellipsoidal algorithm solves some well-known hard LP and LCP problems almost instantly, University of Waterloo Research Report CORR 80-51.

[A7] ———— , On the work of Judin-Nemirovskii on ellipsoidal algorithm and a proof of Hachiyan's result, University of Waterloo Research Report CORR 81-8.

[A8] ———, Solving convex-concave games via ellipsoidal algorithm, University of Waterloo Research Report CORR 81-16.

[A9] ———, On convex programming via ellipsoidal algorithm, University of Waterloo Research Report CORR 81-17.

[A10] ———, Ellipsoidal algorithm: Hachiyan's proof, manuscript, (June 1981), distributed at the talks as a part of Waterloo Rolling Complexity Revue at University of Rochester, and Syracuse.

[A11] ———, Complexity of the ellipsoidal algorithm for a class of linear programs, University of Waterloo Research Report CORR 81-24.

[A12] ———, Scaling: A general framework, (1982) to appear in Math. Operationsforch and Statistics.

[A13] ———, X-matrix simplex method of Makarenkov, 1983.

[A14] ———, An algorithmic proof of the polyhedral decomposition theorem, 1983.

[A15] Ja.I. Al'ber, S.V. Shilman, A unified approach to the problem of minimizing smooth and nonsmooth functions. Engineering Cybernetics 20, No.1 (1982), 21-27.

[A16] D.N. de G. Allen, Relaxation Methods in Engineering and Science, (McGraw-Hill, N.Y., 1954).

[A17] R.E. Anstee, A polynomial algorithm for b-matchings: an alternative approach; Research Report CORR 83-22, University of Waterloo, 1983.

[A18] A. Auslender, Convex programming with errors: methods of ε-subgradients, Survey of Mathematical Programming. (Proc.Ninth International Math. Programming Sympos., Budapest 1976) vol.1 pp. 135-151 (North-Holland, Amsterdam, 1979) MR 83d. 90162.

[A19] B. Aspvall, G. Shiloach, A polynomial algorithms for solving system of linear inequalities with two variables per inequality, SIAM J. Computing 9 (1980), 827-845.

[A20] B. Aspvall and R.E. Stone, Khachiyan's linear programming algorithm, Journal of Algorithms 1, (1980), 1-13. MR81 i. 90123.

[A21] D. Avis, V. Chvátal, Notes on Bland's pivoting Rule, MP Study 8 (1978), 24-38.

[B1] C.V. Babikov, Some relations and pivoting in linear programming, in Methods of Optimization and Pattern Recognition in Problems of Planning, (Russian), pp.37-43 (Akad. Nauk SSSR Ural. Nauchn. Tsentr, Sverdlovsk, 1980). MR 83h. 90052.

[B2] A. Bachem and M. Grötchel, New aspects of polyhedral theory in [K21]. 51-106, MR 83h. 90053.

[B3] A. Bachem, M. Grötchel and B. Korte (eds.), Mathematical Programming: The State of Art, Bonn 1982 (Springer, N.Y., 1983).

[B4] M.L. Balinski, R.E. Gomory, A primal method for the assignment and transportation problems. Management Science 10, 578-594 (1964).

[B5] M.L. Balinski, Signature methods for the assignment problem, Technical Report, Lab. d'Econ. l'ecole polytechniqué Paris, 1982.

[B6] G.M. Bakan, Filtration of multivalued process of bounded noise, Kibernetika i Vych. Teknika 49 (1980), 7-18.

[B7] G.M. Bakan, and E.A. Nizhnichenko, A finite-convergent algorithm for the solution of a countable system of inequalities, Kibernetika Vych. Teknika 53, (1981) 78-83 (Russian), zbl 493.90053.

[B8] R.S. Barr, F. Glover, D. Klingman, The alternating basis algorithm for assignment problems, MP 13 (1977) 1-13.

[B9] R.H. Bartels, A penalty linear programming method using reduced-gradient basis-exchange techniques, Linear Algebra and its Applications 29 (1980), 17-32.

[B10] R.R. Barton, Nondifferentiable optimization methods for a traffic assignment decomposition model, Report 81-EM-018, RCA Labs., Princeton N.J.

[B11] M.S. Bazaara, H.D. Sherali, On the choice of the step size in subgradient optimization, European J. O.R. 7 (1981), 380-388.

[B12] E.N. Belov, Solving linear and nonlinear programming problems by means of the conjugate gradient method together with the penalty prices method, in; Routines and Algorithms 16 (1976) (CEMI, Academy of Sciences SSSR), (Russian).

[B13] A. Ben-Israel, and A. Charnes, An explicit solution of a special class of linear programming problems, O.R. 16 (1968), 1167-1175.

[B14] A. Ben-Israel, Application of generalized inverses to programming, games and networks, in Generalized Inverses and Applications, ed. M.Z. Nashed, (Academic Press, N.Y., 1976).

[B15] A. Ben-Israel, and T.N.E. Greville, Generalized Inverses and its Applications (Wiley, N.Y., 1974).

[B16] G.C. Berresford, A.M. Rockett and J.C. Stevenson, Shachiyan's Algorithm I. A new solution to linear programming problems. BYTE 5 (1980), 198-208.

[B17] ——,——,——, Khachiyan's Algorithm II. Problems with the algorithm, BYTE 5 (1980), 242-255.

[B18] M.G. Bijlani, Dissertation, Clemson University South Carolina, 1983.

[B19] J.R. Birge, and A. Gana, Computational complexity of van der Heyden's variable dimension algorithm and Dantzig-Cottle's principal pivoting method for solving LCP's. MP 26 (1983), 316-325.

[B20] F. Bisshopp, Khachiyan's algorithm for linear inequalities: optimization and implementation, Quart. Appl.Math. 38 (1980), 415-426, MR 83 c. 90087.

[B21] C. Blair, Some linear programs requiring many pivots, University of Illinois, BEBR working paper 867, (Urbana, 1982).

[B22] R.B. Bland, A combinatorial abstraction of linear programming, JCT (B) 23 (1977), 33-57.

[B23] ——————, New finite pivoting rules for the simplex method, MOR 2 (1977) 103-107.

[B24] ——————, D. Goldfarb, and M.J. Todd, The ellipsoid method: a survey, O.R. 29 (1981), 1039-1091 MR 83e.90081.

[B25] ——————, J. Edmonds, Fast primal algorithms for totally unimodular linear programming, XI. International Symp. on Mathematical Programming Bonn, 1982.

[B26] K.H. Borgwardt, The average number of pivot steps required by the simplex method is polynomial. Zeitschrift für Operations Research, A 26 (1982), 157-178.

[B27] K.H. Borgwardt, Some distribution-independent results about the asymptotic order of the average number of pivot steps of the simplex method. M O.R. 7 (1982), 441-462.

[B28] L.M. Bregman, The method of successive projection for finding a common point of convex sets, SMD 6 (1965), 688-692.

[B29] ——————, The relaxation method of finding the common point of convex sets and its application to solution of problems in convex programming. CMMP 7, 3 (1967), 200-217.

[B30] G.W. Brown, and J. von Neumann, Solutions of games by differential equations, Annals of Mathematics Study 24 (1950), 73-79.

[B31] ——————, and T.C. Koopmans, Computational suggestions for maximizing a linear function subject to linear inequalities, in Activity Analysis of Production and Allocation, ed. by T.C. Koopmans (John Wiley, N.Y. 1951), 377-380.

[B32] ——————, Iterative solutions of games by fictitions play, in Activity Analysis of Production and Allocation ed. by T.C. Koopmans, (John Wiley, N.Y. 1951), 376-377.

[B33] V.A. Bulavskii, An iterative method of solution of the problem of linear programming, SMD 2 (1961), 250-252.

[B34] ——————, Relaxation in problems with inequalities, Optimizacija 23 (40), (1979), 32-40, 128-129. (Russian) MR 81h. 90122.

[B35] R.E. Burkard, Subgradient methods in combinatorial optimization, in H. Röck, ed., Discrete Structures and Algorithms, pp.141-151, (Hanser, Munich, 1980), MR 82k. 90083.

[C1] G. Calvillo, Optimum branching systems, Dissertation, University of Waterloo, (1978).

[C2] R. Chadrasekaran, S.N. Kabadi, and K.G. Murty, Some NP-complete problems in linear programming, O.R. Letters 1 (1982), 101-104.

[C3] ——————————, The weighted euclidean 1-center problem, O.R. Letters 1 (1982), 111-112.

[C4] P.M. Camerini, L. Fratta and F. Maffioli, On improving relaxation methods by modified gradient techniques, MP Study 3 (1975), 26-34.

[C5] L. Carleson, Hachian's Method in Linear Programming Report no.3, Institute Mittag-Leffler (1980).

[C6] Y. Cencor, Row action methods for huge and sparse systems and their applications, SIAM Review 23 (1981), 444-466.

[C7] ——————, and A. Lent, Cyclic subgradient projections, MP 24 (1982) 232-235.

[C8] ——————, ——————, An iterative row action method for interval convex programming, JOTA 34 (1981) 321-358.

[C9] A. Charnes, Optimality and degeneracy in linear programming, Econometrica 20 (1952), 160-170.

[C10] ——————, and K. Kortanek, An opposite sign algorithm for 'purification' to an extreme point, ONR Research Report 84, Systems Research Group, Northwestern University (1963).

[C11] ——————, ——————, and W. Raike, An implementation of the opposite sign purification algorithm, Northwestern University, System Research memo 104 (1963).

[C12] ——————, D. Granot and F. Granot, A primal algorithm for interval programming, Linear Algebra and its Applications 17 (1977), 65-78.

[C13] A. Charnes, and W.W. Cooper, S. Duffuas, and M. Kress, Complexity and computability of solutions to linear programming systems, Intern.Jr. of Computer and Information 9 (1980), 483-506.

[C14] V. Chvátal, On certain polytopes associated with graphs, J.C.T. B 18 (1975), 138-154.

[C15] E.W. Cheney and A.A. Goldstein, Newton's Method for convex programming and Tchebycheff approximation, Nemurische Mathematik 1 (1959), 253-268.

[C16] C. Cheng, Iterative methods for solving linear complementarity and linear programming problem, Ph.D. thesis, Computer Science Department, University of Wisconsin, Madison, 1981.

[C17] F.L. Chernousko, Guaranteed estimates of undetermined quantities by means of ellipsoids, SMD 21 (1980), 396-399.

[C18] ———— , Optimal guaranteed estimates of indeterminancies with the aid of ellipsoids, I, II, III, Engineering Cybernetics, 18, no.3 (1980), 1-9, 18, no.4 (1980), 1-9, 18 no.5 (1980) 1-7.

[C19] N.D. Chepurnoi, The relaxation model for the minimization of convex functions, Dokl.Akam.Nauk. Ukr. SSR. A3 (1982) 68-69 (Russian), MR 83d. 90176.

[C20] S.N. Chernikov, Algebraic theory of linear inequalities, Ukranian Mat. z.19 (1967, n.1 36-80). = AMS translations, Ser.2, vol.69 (1968),147-203.

[C21] ————, Linear Inequalities (Russian, Nauka, Moscow, 1968), [Lineare Ungleichunger, (UEB Deutsher, Berlin, 1971)].

[C22] S.J. Chung and K.G. Murty, A polynomially bounded ellipsoid algorithms for convex quadratic programming, Nonlinear Programming 4 (Academic Press, N.Y., 1981), 439-485.

[C23] F.H. Clarke, Generalized gradients and applications, Transactions AMS 205 (1975), 247-262.

[C24] ———, Nonsmooth Analysis and Optimization, (Wiley, N.Y., 1983).

[C25] R.E. Cline, L.D. Pyle, The generalized inverse in linear programming: an intersection projection method and the solution of a class of structured linear programming problems, SIAM Jr.Applied Mathematics 24 (1973) 338-351.

[C26] A. Cobnam, The intrinsic computational difficulty of functions, in Y. Bar-Hillel (ed.) Proc.1964 International Congress for Logic, Methodology, and Philosophy of Science, (North-Holland, Amsterdam 1965), 24-30.

[C27] A.R.Conn, Constrained optimization using a non-differentiable penalty function, SIAM Jr. of Numerical Analysis 10 (1973), 760-874.

[C28] ———, Linear programming via a nondifferentiable penalty function, SIAM Jr. of Numerical Analysis 13 (1976), 145-154.

[C29] S.A. Cook, The complexity of theorem-proving procedures, Proc.ACM Symp. Theory of Computing 3 (1971), 151-158.

[C30] L. Cooper and J. Kennington, Nonextreme point solution strategies for linear programs. Naval Res.Log. Quarterly 26 (1979), 447-461.

[C31] N.E. Cortey, A new method for linear constrained optimization, J. Franklin Institute 313 (1982), 77-83.

[C32] R.W. Cottle and G.B. Dantzig, Complementary pivot Theory of mathematical programming, Linear Algebra and its Applications 1 (1968), 103-125.

[C33] ———, and J.S. Pang, On solving linear complementarity problem as linear programs, MP Study 7 (1978), 88-107.

[C34] R.W. Cottle, J.S. Pang, On the convergence of a block successive over-relaxation method for a class of linear complementarity problems, MP Study 17 (1982), 126-138, zbl 478.90069.

[C35] H. Crowder, Computational improvements for subgradient optimization, Symposia Mathematica, XIX (1976) 357-372.

[C36] C.W. Cryer, The solution of a quadratic programming problem using systematic overrelaxation, SIAM Control 9 (1971), 385-392.

[C37] W.H. Cunningham, Theoretical properties of the network simplex method, M.O.R.4(1979), 196-208.

[C38] ———— , and A. Frank, A primal-dual algorithm for submodular flows, Report no.82224-OR, (1982), Inst.Ökon. Oper.Res., Universität Bonn.

[C39] ———— , Testing membership in matroid polyhedra, Universität Bonn, Inst. für Operations Research Report no.81207-OR (1981).

[C40] ——— , W. Pulleyblank, A.B. Marsh, A scaling algorithm for optimum matching, (1975).

[D1] J.M. Danskin, Fictitions play for continuous game, Naval Res.Log.Quarterly 1 (1955), 313-320.

[D2] ——— , The Theory of Max-min, (Springer-Verlag, N.Y. 1967).

[D3] ——— , The derivative game in max-min and the solution of concave-convex games. Mathematics Research Center TR 586, University of Wisconsin, 1973.

[D4] ——— , On the approximate solution of systems of linear inequalities, (1980.

[D5] ——— , On the approximate solution of concave-convex games over simplexes, (1980).

[D6] G.B. Dantzig, Linear Programming and Extensions, (Princeton University Press, Princeton, N.J. 1963).

[D7] G.B. Dantzig, and P. Wolfe, The decomposition algorithm for linear programming, Econometrica 29 (1961) 767-778.

[D8] ———— , Comments on Khachian's algorithm for linear programming, Technical Report SOL 79-22, Stanford University, 1979.

[D9] ———— , Expected number of steps of the simplex method for a linear program with a convexity constraint, Technical Report SOL 80-3R, Stanford University 1980.

[D10] ———— , Reminisces about the origins of linear programming in [B3], 78-86.

[D11] ———— , M.A.H. Dempster and M. Kallio, eds. Large-scale linear programming (IIASA, Laxenburg, Austria, 1981).

[D12] I. Deak, The use of the ellipsoid method for computing the values of the multivariate normal distribution function, Alkalmaz.Mat.Lapok 2 (1976) 341-349 (Hungarian) MR 57.18019.

[D13] Ph. Delsarte, A refined version of Khachian's algorithm Philips J.Res. 35 (1980), 307-319. MR 81k. 90056.

[D14] V.F. Demjanov, Successive approximation for finding saddle points, SMD 8 (1967).

[D15] ———— , On the solution of certain minimax problems I, Cybernetics 6 (1966).

[D16] ———— , U.N. Malozemov, Introduction to Minimax, (Wiley, N.Y. 1974).

[D17] ———— , Finding saddle points on polyhedra, SMD 11 (1970), 558-561.

[D18] ———— , Subgradient method and saddle points Vestnik Leningrad Univ.Mat.Meth.Astronom. 2(1981), 17-23, MR 83d. 90164.

[D19] D.A. D'esopo, A convex programming procedure, Naval Res. Log.Quarterly 6 (1959), 33-42.

[D20] D.P. Dobkin, S.P. Reiss, The complexity of linear programming, Theoretical Computer Science 11 (1980), 1-18, MR 81e. 90055.

[D21] Z. Drezner, The nested ball principle for the relaxation method. O.R. 31 (1983), 587-590.

[D22] M.E. Dyer, The complexity of vertex enumeration methods, M.O.R. 8 (1983), 381-402.

[E1] B.C. Eaves and W.I. Zangwill, Generalized cutting plane algorithms, SIAM Control 9 (1971), 529-542.

[E2] J.G. Ecker and M. Kupferschmid, An implementation and generalization of Shor's ellipsoid algorithm for convex programming. MP 27 (1983) 83-106.

[E3] J. Edmonds, Paths, trees, and flowers, Can.J.Math. 17 (1965), 449-467.

[E4] ———— , Minimum partition of a matroid into independent subsets, J.Res.Bur.Standards 69B (1965), 67-72.

[E5] ———— , Matroid partition in Mathematics of the Decision Sciences, G.B. Dantzig and A.F.Veinott Jr. (eds.) (American Mathematical Society, Providence, Rhode Island, 1968).

[E6] ———— , Optimum branchings. J.Res.Bur.Standards,71B (1967), 233-240.

[E7] ———— , Maximum matching and a polyhedron with 0-1 vertices. J.Res.Bur.Standards 69E (1965), 125-130.

[E8] ———— , Systems of district representatives and linear algebra, J.Res.Nat.Bur.Standards. 71B (1967), 241-245.

[E9] ———— , and R.M. Karp, Theoretical improvements in algorithmic efficiency for network flow problems, J. ACM 19 (1972), 248-264.

[E10] H.G. Eggleston, Convexity, (Cambridge University Press, Cambridge, 1958).

[E11] J. Elzinga and T.G. Moore, A central cutting plane algorithm for the convex programming problem. MP 8 (1975), 134-145.

[E12] I.I. Eremin, Incompatible systems of linear inequalities, SMD 2 (1961), 821-824.

[E13] ———, Iteration method for Chebyshev approximations for sets of compatible linear inequalities, SMD 3 (1962), 570-572.

[E14] ———, The relaxation method of solving systems of inequalities with convex functions on the left side, SMD 6 (1965), 219-221.

[E15] ———, On systems of inequalities with convex functions in the left sides, Izv. Akad. Nauk SSSR. Ser.Math. 30 (1966), 265-278, = Amer.Math. Society Translations (2) vol.88 (1970), 67-83.

[E16] ———, Penalty method in convex programming, SMD 8 (1967), 459-462.

[E17] ———, Methods of Fejer approximations in convex programming, (Ph.D.thesis) Mathematical Notes 3 (1968), 139-149.

[E18] ———, On the speed of convergence in the method of Fejer approximation, Mathematical Notes 4 (1968), 522-527.

[E19] ———, The application of the method of Fejer approximation to the solution of problems of convex programming with non-smooth constraints, CMMP 9 (1966), 225-235.

[E20] ———, and V.D. Mazuzov, Iterative method for solving problems of convex programming, Soviet Physics Doklady 11 (1967), 757-759.

[E21] ———— , Iteration processes generated by nonstationary Fejér mappings. Trudy Inst.Mat. i Meh.Ural.Naucn.Centr Akad.Nauk SSSR No. 29 Metody dija Nestacionar. Zadac Mat. Programmir. (1979), 3-13, 141. (Russian), MR 82k. 90075.

[E22] ———— , Nonstationary processes of mathematical programming. Methods of optimization and pattern recognition in problems of planning, (Russian), pp.13-19, 147, Akad.Nauk SSSR, Ural, Naucn.Centr, Sverdlovsk (1980), (Russian) MR 82j. 90059.

[E23] ———— , Fejér mappings and nonsmooth optimization. Trudy Inst.Mat. i Meh.Ural.Naucn. Centr. Akad. Nauk SSSR. No.27 Metody Mat.Programmir. i Prilozen (1979), 10-14, 125 (Russian) MR 82d: 90091.

[E24] J.M. Ermolev, Methods of solution of nonlinear extremal problems, Cybernetics 2(4) (1966), 1-14 .

[E25] ———— , N.Z. Shor, A method of random search for a two step problem of stochastic programming and its generalization, Cybernetics 4 (1968), 59-60 MR 45.8265.

[E26] ———— , ————, On the minimization of nondifferentiable functions, Cybernetics 3 (1967), p.72.

[E27] ———— , On convergence of quasi-Fejér sequences, Cybernetics 7 (1971), 655-656.

[F1] T.L. Fam , On multistep methods of linear programming. Dok.Akad.Nauk BSSR 25 (1981), no.9, 784-787, 859. (Russian) MR 83e. 90084.

[F2] ———, The two-step problem with decomposition of support for solving large-scale problems of linear programming. Vestsî Akad.Nauk BSSR Ser.Fiz.-Mat.Nauk 1981, no.4, 45-49. (Russian) MR 83e. 90085.

[F3] ———, Block method of solution of the general linear programming problem. Vetsî Akad.Nauk BSSR Ser.Fiz.-Mat.Nauk 1982, no.2, 44-47, (Russian) MR 83e. 90086.

[F4] ———, The dual two-step method of linear programming. Dokl.Akad.Nauk BSSR 26 (1982), 297-299, (Russian), zbl 494.90049.

[F5] Y. Fathi, Computational complexity of LCP's associates with positive definite symmetric matrices, MP 17 (1979), 335-344.

[F6] M.L. Fisher, Lagrangian relaxation method for solving integer programming problems. Management Sci. 27 (1981), 1-18.

[F7] ———, An equivalence between the subgradient and ellipsoidal algorithm, in [D 11], 479-488.

[F8] R. Fourer, Notes on 'semi-linear' programming, Northwestern University, Dept. of Industrial Eng. and Management Science (1981).

[F9] J.B.J. Fourier, Solution d'uns question particulière du calcul des ineqalités, (1826), Oeuvres II, 317-328.

[F10] A. Frank, How to make a digraph strongly connected, Combinatorica 1 (1981), 145-153.

[F11] K. Fukuda, Oriented matroid programming, Ph.D. thesis University of Waterloo, 1982.

[F12] D.R. Fulkerson, Blocking polyhedra, in Graph Theory and its Applications, ed. B.Harris, (Academic Press, N.Y. 1970).

[F13] ———— , Packing rooted directed cuts in a weighted directed graph, MP 6 (1974), 1-13.

[F14] ———— , G.L. Nemhauser and L.G. Trotter, Two computationally difficult set covering problems that arise in computing the 1- width of incidence matrices of Steiner triple systems. MP Study 2 (1972), 137-163.

[G1] R. Gabasov, V.S. Glunshenkov, A finite exact μ-additive method for solving a general linear programming problem. Dokl.Akad.Nauk BSSR 26 (1982), 589-592, MR 83j. 90053.

[G2] ———— and P.M.Kirillova, The construction of sub-optimal plans of a transportation problem, Engineering Cybernetics 13, no.6 (1975), 34-38.

[G3] ————, ———— , Linear Programming Methods II: General Problems. (Izdatel'stvo Belorusskogo Universiteta, Minsk, 1977) (Russian) zbl 468.90035.

[G4] ————, ———— , Linear programming methods I: Transportation Problems. (Izdatel'stvo Belorusskogo Universiteta, Minsk, 1978), (Russian) zbl 468. 90036.

[G5] ————, ———— , Linear programming methods III: Special Problems. (Izdatel'stvo Belorusskogo Universiteta, Minsk, 1980), (Russian) zbl 484. 90067, MR 82k.90077.

[G6] ————, ———— , and O.I. Kostyukova, Approximate methods for solving the general linear programming problem, Dokl.Akad.Nauk BSSR 24 (1980) 680-683 (Russian) Mr 81k. 90057.

[G7] ————, ———— , A finite direct exact algorithm of linear programming, Dokl.Akad.Nauk BSSR 25 (1981), 301-303 (Russian) zbl 482.90060.

[G8] ———, ———, ———, A finite dual exact algorithm for linear programming. Dokl.Akad.Nauk .BSSR 25 (1981), 586-589 (Russian) MR 83e. 90087.

[G9] ———, ———, ———, Modification of an adaptive method for solving a general problem of linear programming. Vestn.Beloruss.Gos.Univ.Im. V.I. Lenina, Ser.I, 2(1981) 37-41 (Russian) zbl 494. 90048.

[G10] ———, ———, ———, A direct exact algorithm for solving a linear optimal-control problem. Engineering Cybernetics 20, no.1 (1982), 50-56.

[G11] ——— and V.S. Glushenkov, Finite μ-adaptive approximate method for the solution of a general problem of linear programming. Dokl.Akad.Nauk BSSR 26 (1982) 589-592.zbl 494.90045.

[G12] ———, F.M. Kirillova, B.S. Murdakhovich, The ε-maximum principle for suboptimal controls. SMD 27 (1983).

[G13] P. Gács, L. Lovász, Khachiyan's algorithm for linear programming, MP Study 14 (1981), 61-68. MR 83e. 90088a.

[G14] ———, ———, Khachiyan's algorithm for linear programming. Yingyong Shuxue yu Jisuan Shuxue (1980) 1-4 (Chinese) MR. 90088b.

[G15] M.R. Garey and D.S. Johnson, Computers and Intractability: a Guide to the Theory of NP-completeness. (Freeman, San Francisco, 1979).

[G16] R.B. Gasber, Ein Erfahrungsbericht über die implementation eines verbesserten Kachijan-algorithmus. Angew. Inf. 23, 502-505 (1981), zbl 468.90038.

[G17] A.M. Geoffrion, An improved implicit enumeration approach for integer programming. O.R. 17 (1969), 437-454.

[G18] V.I. Gershovich, About a cutting method with linear space transformations. In Theory of Optimal Solutions, (Kiev, Inst. of Cybernetics. Ac.Sci. Ukr.SSR., 1979, pp.15-23 (Russian) MR 82d.90096.

[G19] ———, One way of representing space transformation operators in high-speed versions of generalized gradient methods. In Numerical Methods of Nonlinear Programming. Summaries of Papers at the III. All-Union Seminar.-Kharkov, 1979, pp.64-66 (Russian).

[G20] ———, Efficient algorithm for solving problems of nonsmooth optimization that combines cutting methods and linear transformations in variable space. In Systems of Software Support for Solution of Optimal Planning Problems. Summaries of Papers at the IV. All-Union Symp. M., 1980, pp.50-51, (Russian).

[G21] ———, One optimization method using linear space transformations. In Theory of Optimal Solutions (Kiev, Inst. of Cybernetics, Ac.Sci.Ukr. SSR., 1980), pp.38-45, (Russian).

[G22] ———, About an ellipsoid algorithm. In, Some Algorithms of Nonsmooth Optimization and Discrete Programming. Preprint 81-6 Inst.of Cybernetics, Ac.Sci.Ukr.SSR., 1981, pp.8-13, (Russian).

[G23] ———, Quadratic smoothing in iterative decomposition algorithms of solution of large-scale linear-programming problems. In II. Republican Symp. on Methods of Solution of Nonlinear Equations and Optimization Problems. (Tallin, 1981, pp.188-190), (Russian).

[G24] P.E. Gill, W. Murray and M.A. Saunders, Methods for computing and modifying the LDV factors of a matrix, Mathematics of Computation 29 (1975), 1051-1077.

[G25] ———, ———, ———, M.H. Wright, A numerical investigation of ellipsoid algorithms for large-scale linear programming, in [D11] 487-511.

[G26] J.L. Goffin, On convergence rates of subgradient optimization methods, MP, pp.329-347, 13 (1977).

[G27] ———, Nondifferentiable optimization and the relaxation method, in Nonsmooth Optimization, proceedings of a IIASA Workshop, 28 March-9 Apr., 1977 (pp.31-50), Eds.: C.Lemarechal and R.Mifflin, IIASA Proceedings Series (1978) Pergamon Press. MR 81f.90092.

[G28] ———, On the suprising effectiveness of subgradient optimization. Presented at the 10th Intern. Mathematical Programming Symp. in Montréal,1979.

[G29] ———, The relaxation method for solving systems of linear inequalities, M O.R. 5, pp.388-414 (1980). MR 83a. 90086.

[G30] ———, Convergence results in a class of variable metric subgradient methods, in Nonlinear Programming 4, Editors: O.L.Mangasarian, R.R. Meyer and S.M. Robinson, Academic Press (1981), pp. 283-326.

[G31] ———, Chaos and order in relaxation methods with maximizing controls, in proceedings of the 6th Conference on Probability Theory (Brasov, Romania, Sept.10-15, 1979), Editors: B.Bereanu, S. Grigorescu, M. Iosifescu and T. Postelnicu, pp.67-76 (1981) Editura Academiei Republicii Socialiste Romania, Bucuresti, Romania.

[G32] ———, On the nonpolynomiality of the relaxation method for systems of linear inequalities, MP 22, (1982), pp.93-103. MR 83c. 90090.

[G33] ———, Convergence of a cyclic ellipsoid algorithm for systems of linear equalities, MP 22, (1982) pp.239-260, MR.83d. 90115.

[G34] ———, Acceleration in the relaxation method for linear inequalities and subgradient optimization, in Progress in Nondifferentiable Optimization, Ed., E.A. Nurminski, IIASA Collaborative Proceedings Series, CP-82-S8, IIASA, Laxenburg, Austria (1982), pp.29-59.

[G35] ———, Convergence rates of the ellipsoid method on general convex functions, M.O.R. 8, (1983) p.235-250.

[G36] ——— and A.J. Hoffman, On the relationship between the Hausdorff distance and matric distances of ellipsoids, Linear Algebra and its Applications, 52/53, pp. 301-313 (1983).

[G37] ———, Variable metric relaxation methods, part I: a conceptual algorithm, SOL Technical Report 81-16, Dept. of Operations Res., Stanford Univ. California, U.S.A.; pp.101, (Aug.1981).

[G38] ———, Variable metric relaxation methods, part II: an implementable algorithm, or the ellipsoid method, Faculty of Management, McGill Univ., Montréal, Quebec, Canada, Working Paper No.82-27, p.29, (Aug.1982).

[G39] D. Goldfarb and W.Y. Sit, Worst case behavior of the steepest edge simplex method, Discrete Applied Math.1 (1979), pp.277-285.

[G40] ——— and M.J. Todd, Modifications and implementation of the Shor-Khachian algorithm for linear programming, MP.23 (1982), 1-19, MR 83d: 90116, zbl 477.90038.

[G41] V. M. Goldman, Einige Erläuterungen zum Khacian-Verfahren Zeitschrift für O.R. B 25(1981), 95-112.

[G42] A.A. Goldstein, Optimization of Lipschitz, Continuous functions, MP.13 (1977), 14-22.

[G43] E.G. Golhstein, Generalized gradient method for finding saddle points, EMM 8 (1970) = MATEKON.

[G44] ———, An iterative linear programming algorithm based on an augmented lagrangian, Nonlinear Programming 4 (Academic Press, N.Y., 1981), 131-146, also in [D] MR 83d: 90055.

[G45] R.E. Gomory, An algorithm for integer solutions to linear programs, in R.L. Graves and P.Wolfe eds. Recent Advances in Mathematical Programming, (McGraw-Hill, New York, 1963).

[G46] R. Gorenflo and Y. Kovetz, Solution of an Abel-type integral equation in the presence of noise by quadratic programming, Numer.Math., 8 (1966), pp.392-406.

[G47] M. Grötschel, W.R. Pulleyblank, Weakly bipartite graphs and the max-cut problem, O.R. Letters 1 (1981) 23-27, MR 83e. 05048.

[G48] ———, L. Lovász and A. Schrijver, The ellipsoid method and its consequence in combinatorial optimization, Combinatorica 1 (1981), 169-197.

[G49] ———, ———, ———, Polynomial algorithms for perfect graphs, Annals of Discrete Mathematics (to appear).

[G50] B. Grünbaum, Convex Polytopes (Wiley, N.Y., 1967).

[G51] L. Gubin, B. Poljak, E. Raik, The method of projections for finding the common point of convex sets. CMMP 7 (1967).

[G52] H. Guggenheimer, Applicable Geometry (Krieger, N.Y., 1977).

[G53] E.A. Gunn and G.J. Anders, A comparison of interval linear programming with the simplex method. Linear Algebra and its Applications 38 (1981) 149-159.

[G54] J. Gwinner, Bibliography on non-differentiable optimization and non-smooth analysis. Jr. of Computational and Applied Math. 7 (1981) 277-285.

[H1] L.G. Hachiyan, Convergence rate of the game process for solving matrix games, CMMP 17 (1977), 78-88.

[H2] ———, A polynomial algorithm in linear programming, SMD 20 (1979), 191-194, MR 80g. 90071.

[H3] ———, A polynomial algorithm in linear programming, Yingyong Shuxue yu Jisuan Shuxue 2 (1980) 1-3 (Chinese), MR 83h. 90058.

[H4] ———, Polynomial algorithms in linear programming, CMMP 20 (1980), 53-72.

[H5] ———, On exact solutions of systems of linear inequalities and LP problems, CMMP 22 (1982) 999-1003.

[H6] S. Halfin, The sphere method for Khachiyan's algorithm, MP (1983), 109-116.

[H7] G.H. Hardy and E.M. Wright, An Introduction to the Theory of Numbers, (Oxford Press, 1965).

[H8] J. Hartmanis and R.E. Stearns, On the computational complexity of algorithms, Notes for University of Michigan Summer Conference on Automata Theory 1963, 59-79, also in Transactions A.M.S. 117 (1965), 285-306.

[H9] M. Held and R.M. Karp, The traveling-salesman problem and minimum spanning trees, Part II, MP. 1 (1971), 6-25.

[H10] ———, ———, and P. Wolfe, Large scale optimization and the relaxation method, in Proc. of the 25th Conference of the ACM (1972), 507-509.

[H11] ———, P. Wolfe and H. Crowder, Validation of subgradient optimization, MP. 6 (1974), 62-88.

[H12] R. Helgason, J. Kennigton, and H. Lall, A polynomially bounded algorithm for a singly constrained quadratic program, MP.18 (1980), 338-343.

[H13] M. Hendrikse-Baardman, Alternatives for the simplex method, M.S. Thesis, Econometric Institute, Erasmus University, Rotterdam (1980), (Dutch).

[H14] G.T. Herman, A. Lent, Iterative reconstruction algorithms, Comput. Biol. Med. 6 (1976), 273-294.

[H15] C. Hildreth, A quadratic programming procedure, Naval Res.Log.Quart., 4 (1957), pp.79-85.

[H16] A.Y. Hinchin, Continued Fractions, (The University of Chicago Press, Chicago, 1964).

[H17] J.B. Hiriart-Urruty, ε-subdifferential calculus, in Convex Analysis and Optimization. Eds. J.P. Aubin and R.B. Vinters, (Pitman Research Notes in Math. 58, 1981).

[H18] A.J. Hoffman, On approximate solutions of systems of linear inequalities, J.Res.Nath.Bur.Standards 49 (1952), 263-265.

[H19] ———, and J.B. Kruskal, Integral boundary points of convex polyhedra, in Linear Inequalities and Related Systems, H.W. Kuhn and A.W. Tucker (eds.), (Princeton University Press, Princeton, N.J. 1956).

[H20] ———, On the use of unimodularity in applying linear inequalities to combinatorial problems, Annals of Discrete Math. 4 (1979), 73-84.

[H21] R. Horst, A note on functions whose local minima are global, J.O.T.A. 36 (1982), 457-462.

[H22] M.S. Hung, A polynomial simplex method for the assignment problem. O.R. 31 (1983), 595-601.

[I1] Y. Ikura and G.L. Nemhauser, An efficient primal simplex algorithm for maximum weighted vertex packing on bipartite graphs, Annals of Discrete Mathematics 16 (1982), 149-168.

[I2] A.D. Ioffe , An approach to nonsmooth analysis, Transactions of AMS 266 (1981), 1-56.

[I3] V.V. Ivanov, Optimal algorithms of minimization of certain classes of functions, Cybernetics 4 (1972), 620-634.

[I4] ———, V.A. Ljudvichenko, V.S. Mihalevich, V.A. Trubin, N.Z. Shor, Questions of increasing the effectiveness of algorithms for minimization of functions and mathematical programming, Akad.Nauk. USSR Cybernetics Institute, preprint, 1979 (Russian), MR 91i. 90147.

[I5] L.V. Ivonova, N.Z. Shor, A certain iterative method of solution of linear programming problems and matrix games in, Theory of Optimal Solutions 3 (Kiev, 1968), 22-30, (Russian) MR 46.4930.

[J1] R.G. Jeroslow, Some relaxation methods for linear inequalities, Cahiers du C.E.R.O. 21 (1979), 43-53, MR 80g: 49038.

[J2] ———, The simplex algorithm wit the pivot rule of maximizing criterion improvement, Discrete Math. Math. 4 (1973), 367-377.

[J3] F. John, Extremem problems with inequalities as subsidiary conditions, in Courant Anniversary Volume, (Wiley, N.Y., 1948).

[J4] P.C. Jones and E.S. Marwil, A variant of Khachiyan's algorithm for linear programming, Nov.1979, EG and G Idaho, Inc., P.O.B. 1625, Idaho Falls, Idaho 83415.

[J5] ———, ———, Dimensional reduction variant of Khachiyan's algorithm for linear programming Problems, M.O.R. 7 (1982) 245-252. Idaho.

[J6] ———, ———, Solving complementarity problems with Khachiyan's algorithm, Jan.1980, EG and G Idaho Inc., P.O.B. 1625, Idaho Falls, Idaho 83415.

[J7] D.B. Judin and A.S. Nemirovskii, Estimation of the informational complexity of mathematical programming problems, EMM 12 (no.1, 1976), 128-142= MATEKON 13 (No.2, 1977), 3-25. MR 56.8004.

[J8] ———, ———, Informational complexity and effective methods for the solution of convex extremal problems, EMM 12 (no.2, 1976), 357-369 = MATEKON 13 (no.3, 1977), 25-45. MR 56.10965.

[J9] ———, ———, Informational complexity of strict convex programming, EMM 13, (no.3, 1977), 550-559 (Russian). MR 56.7990.

[J10] ———, ———, Efficient methods for solving large scale convex programming problems, EMM 15 (1979) 135-152 = MATEKON 16, no.3 (1980), 23-52 MR 80e. 90083.

[J11] ———, The effectiveness of random search in control problems, Engineering Cybernetics 15, no.3 (1977), 1-14.

[J12] ———,———, Effectiveness of randomization of control, Plomlemy Sluchain Poiska no.7 (1978), 22-66. (Russian), MR 80e. 93145.

[J13] ———, Mathematical rogramming in order scales, Engineering Cybernetics 20, no.2 (1982), 1-12.

[J14] ———, A.P. Goryashko and A.S. Nemirovskii, Mathematical Methods of Optimization of Algorithms and Devices of Automatic Control Systems, (Moscow, 1982), (Russian).

[K1] S. Kakutani, A generalization of Brouwer's fixed point theorem, Duke Math.J. 8 (1941), 457-459.

[K2] M. Kallio, E.L. Porteus, A class of methods for linear programming, MP 14 (1978), 161-169.

[K3] ———, W. Orchard-Hays, Experiments with the reduced gradient method for general and dynamic linear programming, in [D11] 631-662.

[K4] H.S. Kang, Y.H. Park, The application of Khachiyan's algorithm for linear programming, State of the Art. J. Korean Oper. Res. Soc. 6, 65-70 (1981). zbl, 478.90049.

[K5] L.V. Kantorovitch, A new method of some classes of extremal problems, Comptes Rendus (Doklady) de l'Academie des Sciences de l'URSS (1940) vol. 28, pp.211-214. MR 2.222.

[K6] S. Karamardian, A note on Khatchian's algorithm. J.O. T.A. 32 (1980), 595-597. MR 83e. 90090.

[K7] R.M. Karp, Reducibility among combinatorial problems, in R.E. Miller and J.W. Thatcher (eds.), Complexity of Computer Computations, (Plenum Press, New York 1972).

[K8] ———, On the computational complexity of combinatorial problems, Networks 5 (1975), 45-68.

[K9] ———, and C.H. Papadimitriou, On linear characterizations of combinatorial optimization problems. SIAM Jr.Comp. 11 (1982) 620-632. MR 83k.90064.

[K10] G.M. Karpelevich, The extragradient method for finding saddle points and other problems, EMM 12 (1976), 747-756 = MATEKON 13, Summer 1977, 35-49.

[K11] ———, The method of coordinatewise descent for problems of minimization with constraints of linear inequalities and matrix games, in Math. Methods for the Solution of Econoproblems No.9, 84-93, (Nauka , Moscow, 1980), (Russian) MR 83j. 90080.

[K12] A.V. Karzanov, On the minimal number of arcs of a digraph meeting all its directed cutsets, (to appear).

[K13] J.E. Kelley, Jr. The cutting plane method for solving convex programs, SIAM Appl.Math. 8 (1960), 703-712.

[K14] V. Klee and G.L. Minty, How good is the simplex algorithm?,in O. Shisha (ed.) Inequalities III, (Academic Press, New York, 1972), 159-175.

[K15] J.L. Kiehm and P. Vincke, La method de Khachyan: Description et justification mathematique, Cahiers du C.E.R.O. 23 (1981), 105-120.

[K16] ———, ———, Methode ellipsoidale et algorithme du simplexe: complexite theorique et pratique, Cahiers du C.E.R.O. 24 (1982) 297-304 zbl. 90052.

[K17] A.I. Korablev, O.V. Mironov, An attempt at solution of model problems of minimization by the ε-subgradient method, Issled.Prikl.Math. 8 (1980) 9-14 , MR 82m. 90137.

[K18] ———, ———, An algorithm for the ε-subgradient method of minimization and its application. Issled. Prikl.Math. 7 (1979), 12-18 (Russian). MR 82a. 90142.

[K19] B. Korte and R. Schrader, A note on convergence proofs of Shor-Khachian methods, in Lecture notes in Control and Information Sciences no.30, (Springer, N.Y., 1980) 51-57.

[K20] ———, ———, Can the ellipsoid method be efficient, in Proceedings in Linear Optimization Algorithms and Software, Pisa 1980, (to appear).

[K21] ———, Modern Applied Mathematics, (ed.) (North-Holland, N.Y., 1982).

[K22] M.K. Kozlov, S.P. Tarasov, L.G. Hachiyan, Polynomial solvability of convex quadratic programming, SMD 20, (1979), 1108-1111.

[K23] M.I. Kreinin, A relaxation subgradient algorithm for minimizing convex functions, Soviet Mathematics (Iz.VVZ) 24 (1980), 101-103, MR 81i. 90168.

[K24] Y. Krol, B. Mirvan, Some modifications of ellipsoid method for LP problems, Arcon, Mass. 1980.

[K25] H.W. Kuhn, Pivotal exchange algorithms, in Nonlinear Optimization, eds. L.C. Dixon, E. Spedicato, G.P. Szegö, (Birkhäuser, Boston, 1980).

[K26] A.I. Kuksa, N.Z. Shor, The method of estimating the number of conditionally optimal trajectories of discrete separable dynamic programming. Cybernetics 8 (1972) 910-919, MR 48.1463.

[K27] G.S. Kumar, An approach to linear programming using relaxation methods, Northern Illinois University, 1980.

[K28] M. Kupferschmid, An ellipsoid algorithm for convex programming, Ph.D. Dissertation, Rensselaer Polytechnic Institute, Troy, N.Y., 1981.

[K29] A.T. Kutanov, Refining the solution of the linear programming problem in the method of penalty functions, Automation and Remote Control 6 (1970) 626-632.

[K30] S.S. Kutataledzade, Convex ε-programming, SMD 20, (1979) 391-393.

[K31] ———, ε-subdifferential and ε-optimality, Siberian Mathematical Jr. 21 (1980), 404-411.

[K32] A.I. Kuzevkin, and V.M. Thomirov, On a quantity of observations for finding minimum of a convex function, EMM 3 (1967), 95-103 (Russian).

[L1] L.S. Lasdon, Optimization Theory for Large Systems, (MacMillan, N.Y., 1970).

[L2] E.G. Lawler, Combinatorial Optimization, Networks and Matroids, (Holt, Rinehart and Winston, N.Y. 1974).

[L3] ———, The great mathematical sputnik of 1979, Math.Intellengencer 2 (1980), 191-198 MR 82e. 90054.

[L4] ———, C.U. Martel, Computing maximal polymatroidal network flows, M.O.R. 7 (1982) 334-347.

[L5] ———, ———, Flow network formulations of polymatroid optimization problems. Annals of Discrete Mathematics 16 (1982) 189-200.

[L6] ———, Generalizations of the polymatroid network flow model, preprint BW 158/82, Mathematisch Centrum, Amsterdam, 1982.

[L7] C.W. Lee, Characterizing the number of the faces of a simplical convex polytope, in, Convexity and Related Areas, (Marcel Dekker, N.Y., 1982).

[L8] C. Lemarechal, nonsmooth optimization and descent methods, IIASS, RR 78-4, 1978.

[L9] ———, and R. Mifflin, eds., Nonsmooth optimization, IIASA Proceedings series, Pergamon Press (1978).

[L10] C.E. Lemke, Bimatrix equilibrium points and mathematical programming, Management Science 4 (1965), 681-689.

[L11] A.Ju Levin, An algorithm for the minimization of convex functions, SMD 6 (1965), 286-290.

[L12] L. Lovász, Normal hypergraphs and the perfect graph conjecture, Discrete Math. 2 (1972), 253-267.

[L13] ———, A. Schrijver, Some combinatorial applications of the new linear programming algorithm. In Lecture Notes in Math., 885, (Springer, Berlin, 1981), 33-41, MR 83h. 90071.

[L14] ———, On the Shannon capacity of a graph, IEEE Trans. on Information Theory 25 (1979), 1-7.

[L15] ———, Submodular functions and convexity, in [B3] 235-257.

[L16] ———, Bounding the independence number of a graph, Annals of Discrete Mathematics 16 (1982) 213-223.

[L17] C.L. Luccesi, A minimax equality for directed graphs, Ph.D. thesis, University of Waterloo, 1976.

[L18] D.L. Luenberger, Introduction to Linear and Nonlinear Programming, (Addison Wesley, London, 1973).

[L19] Yu. I. Lyubich and G.D. Maistrovskii, The general theory of relaxation processes for convex functionals, Russian Mathematical Surveys (1970), 57-117.

[M1] Z.F. Ma, F. Tian, A polynomial algorithm for the linear programming problem, Knowledge Practice Math., (1980) 36-44, (Chinese) MR 82c. 90054.

[M2] D. Maistroskii, Gradient methods for finding saddle points, EMM 12 (1976), 917-929, = MATEKON 14, Fall 1977, 3-22.

[M3] Ju. M. Makarenkov, The x-matrix simplex method, SMD. 24 (1981) 24-27, zbl 478.90050.

[M4] U.H. Malkov, G.G. Padchin, N.A. Sokolov, Some techniques to improve the efficiency of solving linear programming problems, in [D11], 699-708.

[M5] J. Mandel, Convergence of relaxation methods for linear inequalities, Technical Report ST24, Charles University, Praha, (1982).

[M6] O.L. Mangasarian, Nonlinear Programming, McGraw-Hill, New York, 1969.

[M7] ———, Characterization of linear complementarity problems as linear programs, MP Study 7, (1978), 74-77.

[M8] ———, Iterative solution of linear program, SIAM Numerical Analysis 18 (1981), 606-614.

[M9] ———, Least norm linear programming solution as an unconstrained minimization problem, Jr.Math.Analys.Applications 92 (1983) 240-251 .

[M10] ———, A condition number for linear inequalities and linear programs, Methods of O.R. 43 (1981) 3-15.

[M11] S.V. Markov, Support methods for solving a multi-product transportation network problem. Problems of Optimal Control, pp.305-318, 375-376, "Nauka i Tekhnika", Minsk, 1981 (Russian) MR 83d. 90131.

[M12] J.F. Maurras, Bon algorithmes, Vieilles ideas, Note E.d.F. Hr 32, 0320, 1978.

[M13] ———, Good algorithms, old ideas, manuscript, 1978.

[M14] ———, K. Truemper and M.Akgül, Polynomial algorithms for a class of linear programs, MP 21 (1981), 121-136.

[M15] E.H. McCall, A study of the Khachian algorithm for real-world linear programming, Tech. Report, University of Minnesota, 1980.

[M16] ———, Performance results of the simplex algorithm for a set of real-world linear programming problems, Technical Report, University of Minnesota, 1980.

[M17] Y.I. Merzlyakov, On a relaxation method of solving systems of linear inequalities, CMMP 2 (1962) 504-510.

[M18] V.S. Mikhalevich, I.V. Sergienko and N.Z. Shor, Investigation of optimization methods and their applications, Cybernetics 17 (1981) 522-548.

[M19] ———, N.Z. Shor and N.G. Zhurbenko, et al., Numerical methods for the selection of optimal design decisions, (Naukova Dumka, Kiev, 1977) (Russian) MR 57.2588.

[M20] N. Megiddo, Is binary encoding appropriate for the problem-language relationship? Theoretical Computer Science 19 (1982) 337-341, MR 83h. 90054.

[M21] ———, Towards a genuinenly polynomial algorithm for linear programming, Computing 12 (1983) 347-353.

[M22] ———, Solving linear programming problems in linear-time when the dimension is fixed. JACM (to appear).

[M23] ———, Linear-time algorithms for linear programming in R^3 and related problems, SIAM Computing (to appear).

[M24] M. Minox and J.Y. Serreault, Subgradient optimization and large scale programming: an application to optimum multicommodity network synthesis with security constraints. R.A.I.R.O. Operations Research 15 (1981) 185-203.

[M25] G.J. Minty, On the axiomatic foundations of the theories of directed linear graphs, electrical networks, and network programming, J.Math. and Mechanics 15 (1966), 485-520.

[M26] B.S. Mitiagin, Two inequalities for volumes of convex bodies, Mathematical Notes, vol.5 (1968), 61-65.

[M27] T. Motzkin and I.J. Schoenberg, The relaxation method for linear inequalities, Canadian Journal of Mathematics 6 (1954), 393-404.

[M28] W. Murray, Ellipsoidal algorithm for linear programming. Working Paper 79-1, Department of Operations Research, Stanford University, Stanford, CA, 1979.

[M29] K.G. Murty, Computational complexity of complementary pivot methods, MP Study 7 (1978), 61-73.

[M30] ———, Linear Programming, (Wiley, N.Y., 1983).

[M31] V.N. Myagkov, The method of horizontal bordering for network problems of the transportation type Optimizatsiya 25 (42) (1980), 85-98 (Russian), zbl 484.90089.

[M32] A.B. Marsh, Matching Algorithms, Ph.D.Thesis, The John Hopkings University, 1979.

[N1] B. Näslund, Hachiyan's algorithm, Normat 29 (1981), 162-169 (Swedish), MR 83e. 90091.

[N2] A.N. Nemirovskii, D.B. Judin, Optimization methods adapting to the 'significant' dimension of the problem, Automation and Remote Control 38, (1977) No.4, 513-524. MR 58.90083.

[N3] ———, ———, Complexity of Problems and Efficiency of Methods of Minimization, (in Russian, Moscow, 1979). English translation to appear (Wiley, N.Y., 1983) MR 82e. 90081.

[N4] ———, ———, Cesari convergence of the gradient method of approximating saddle points of convex-concave functions, SMD 19 (1978) 482-486, MR 58.2560.

[N5] ———, The unit-vector method of smooth convex minimization, Engineering Cybernetics 20, no.2 (1982) 13-23.

[N6] D.J. Newman, Location of the maximum on unimodal surfaces, JACM 12 (1965), 395-398.

[N7] E.A. Nurminski, Progress in Nondifferentiable Optimization, (ed.), IIASS (Laxenburg, Austria, 1982).

[O1] W. Oettli, On iterative method, having linear rate of convergence, for solving a pair of dual linear programs, MP 3 (1972), 302-311.

[O2] ———, Symmetric duality and a convergent subgradient method for discrete, linear, constrained approximation problems with arbitrary norms appearing in the objective function and in the constraints, Jr. Approximation Theory 14 (1975), 43-50.

[O3] I.G. Ovrutskii and N.Z. Shor, Application of methods of minimization of nonsmooth functions to the interpretation of gravimetric observations, Cybernetics 12 (1976), 232-239.

[P1] M.W. Padberg, Characterization of totally unimodular, balanced and perfect matrices in B. Roy (ed.). Combinatorial Programming: Methods and Applications, (Reidel, Boston, 1975).

[P2] ———, M. Rao, Odd minimum cut-sets and b-matchings, M.O.R. 7 (1982), 67-80.

[P3] ——— and M. Rao, The Russian method for linear inequalities and linear optimization, GBA working paper, New York University, N.Y., N.Y., 1980.

[P4] ———, ———, The Russian method for linear inequalities. II Approximate arithmetic, GBA working paper, New York University, N.Y., N.Y., 1980.

[P5] ———, ———, The Russian method and integer programming, to appear MP Study.

[P6] C.H. Papadimitriou, Efficient search for rationals, Inf.Proc.Letters 8, 1 (1979), 1-4.

[P7] ——— and K. Steiglitz, Combinatorial Programming: Algorithms and Complexity. (Wiley, N.Y., 198).

[P8] S.R. Paranjape, The simplex method: two basic variable replacement, Management Science 12 (1965), 135-141.

[P9] P.F. Pickel, Some improvements to Khachiyan's algorithm in linear programming, Dec.1979, pp.26. Polytechnic Institute of New York.

[P10] ———, An ellipsoid method for integer linear programming, 1980.

[P11] B.T. Poljak, Gradient methods for solving equations and inequalities, CMMP 4 (1964), 17-32.

[P12] ———, A general method of solving extremum problems, SMD 8 (1967), 593-597.

[P13] ———, Minimization of unsmooth functionals, CMMP 9 (1969), no.3, 14-29.

[P14] ———, Convergence and convergence rate of iterative stochastic algorithms I. General case, Automation and Remote Control 12 (1976), 1858-1868.

[P15] ———, Subgradient methods: a survey of Soviet research, in [L8].

[P16] ———, U.N. Tretyakov, On an iterative method of linear programming and its economic interpretation EMM 8 (1972) 740-751 (Russian).

[P17] ———, Ja. Z. Cypkin, Optimal pseudogradient algorithms of stochastic optimization, SMD (1980), MR 81i. 90160.

[P18] ———, ———, Optimal pseudogradient adaptation algorithms, Automation and Remote Control 41 (1980), 1101-1110, MR 82h. 90082.

[P19] W. Pulleyblank, Faces of matching polyhedron, Ph.D. thesis, University of Waterloo,(1973).

[P20] ———, Polyhedral Combinatorics, in [B3] 312-345.

[P21] L.D. Pyle, The generalized inverse in linear programming: basic structure, SIAM Applied Mathematics 22 (1972), 335-355.

[P22] ——— and R.E. Cline, The generalized inverse in linear programming - interior gradient projection methods, SIAM Applied Mathematics 24 (1973), 511-534.

[R1] M.O. Rabin, Real time computation, Israel Jr. of Math. 1 (1963), 203-211.

[R2] B. Randol, The ellipsoid method in linear programming Adv. in Appl.Math. 1 (1980), 1-6, MR 82d:90077.

[R3] S.P. Reiss, Rational search, Information Processing Letters, 8, (1979),89-90. MR 80g.68062.

[R4] A.G. Rinnooy Kan and J. Telgen, The complexity of linear programming, _Statistica Neerlandica_ 35 (1981) 91-107, MR 83a. 90096.

[R5] P.D. Robers, A. Ben-Israel, A suboptimization algorithm for interval linear programming, a new method for linear programming. _Linear Algebra and its Applications_ 3 (1970) 383-405.

[R6] J. Robinson, An iterative method of solving a game. _Annals of Mathematics_ 54 (1951), 296-301.

[R7] R. Tyrell Rockafellar, _Convex Analysis_, (Princeton University Press, 1970).

[R8] ———, The elementary vectors of subspace of R^n , in _Combinatorial Mathematics and its Applications_, ed. by R.C. Bose and T.A. Dowling (The Univ. of North Carolina Press, 1969).

[R9] ———, _The Theory of Subgradients and its Applications to Problems of Optimization_ (Heldermann, Berlin

[R10] H. Röck, Scaling techniques for minimal cost network flows, in _Discrete Structures and Algorithms_, (Hanser, Berlin, 1980) 181-191.

[R11] J. Rosen, The gradient projection method for nonlinear programming I. Linear constraints, _SIAM Appl. Math_. 8 (1960) 181-217.

[R12] ———, and G, Frawley, A computational test of Khachian's algorithm for linear inequalities. Technical Report 79-28. Computer Science Dept., University of Minnesota, Minneapolis, MI, 1979.

[R13] S.V. Rzhevskii, An ε-subgradient method for solving convex programming problems. _CMMP_ 21 (1981) 51-57 zbl 495.90068.

[R14] ———, Fundamental algorithm for minimizing convex functions. Cybernetics 15, 787-789 (1980); zbl 495.90069.

[R15] M.R. Rao, K. Truemper, On b-matching algorithm, Technical report, University of Texas at Dallas, 1975.

[S1] A.L. Sadovskii, Monotonic iterative algorithms for solving matrix games. Soviet Physics Doklady 23 (1978), 16-17.

[S2] ———, A game approach to the solution of mathematical programming problems. SMD 20 (1979) 446-449.

[S3] R. Saigal, The fixed point approach to nonlinear programming. MP Study 10 (1979), 142-157.

[S4] S. Schecter, Relaxation methods for convex problems. SIAM Numerical Analysis 5 (1968), 601-612.

[S5] R. Schrader, Ellipsoid methods, in [K21] 265-311.

[S6] P.D. Seymour, Decomposition of regular matroids, JCT B 28 (1980), 305-359.

[S7] H.D. Sherali, A.L. Soyster, and S.G. Baines, Nonadjacent extreme point methods for solving linear programs. Naval Res.Log.Quart. 30 (1983) 145-161.

[S8] C. Sandi, Subgradient optimization, in Combinatorial Optimization, eds. N. Christofides, A.Mingazzi, P. Toth and C. Sandi (Wiley, N.Y., 1979).

[S9] N.Z. Shor, Application of the gradient method for the solution of network transportation, Problems of Cybernetics and Operations Research, Acad. of Sciences, Kiev (1962), Russian.

[S10] ———, On the structure of algorithms for the numerical solutions of optimal planning and design problems. Dissertation, Cybernetics Institute, Academy of Sciences of the Ukrainan SSR, Kiev (1964).

[S11] ———, The rate of convergence of the generalized gradient descent method, Cybernetics 4 (1968), 79-80 MR 45.8265.

[S12] ———, and V.I. Biletskii, The method of space dilatation to accelerate convergence in Gully-type problems, in Optimal Decision Theory (in Russian), (vol.2, Kiev, 1968) MR 45.7952.

[S13] ———, Utilization of the operation of space dilatation in the minimization of convex functions, Cybernetics 6 (1970), 7-15 MR 45.4836.

[S14] ———, Convergence rate of the gradient descent method with dilatation of the space, Cybernetics 6(2) (1970), 102-108. MR 45.6452.

[S15] ———, and N.G. Zhurbenko, A minimization method utilizing the operation of space expansion in the direction of the difference of two successive gradients, Cybernetics 7 (1971), 450-459, MR 46.4949.

[S16] ———, A class of almost-differentiable functions and a minimization method for functions of this class. Cybernetics 8, (1972), 599-606, MR 46.8666.

[S17] ———, Generalized gradient methods of non-differentiable function minimization and their application to problems of mathematical programming, EMM 12 (1976), 337-356. (Russian) MR 56.10010.

[S18] ———, Cut-off method with space extension in convex programming problems, Cybernetics 13 (1) (1977) 94-96, MR 56.14711.

[S19] ———, New development trends in non-differentiable optimization, Cybernetics 13 (6) (1977), 881-886, MR 58.2394.

[S20] ———, V. Gershovich, On one family of solution algorithms for problems of convex programming, Cybernetics 15 (1979), 502-508, MR 81k : 90083.

[S21] ———, Generalized gradient methods of non-differentiable optimization employing space dilatation operations, in [B3] 501-529.

[S22] ———, P.R. Gambord, Certain questions of generalized gradient descent, Cybernetics 7 (1971) 1033-1036. MR 46.3083.

[S23] ———, M.B. Schepakin, Multistage stochastic programming problems in parametric form, in Theory of

Optimal Solutions, no.3 (Kiev, 1968) (Russian) MR 46.3078.

[S24] ———, ———, An algorithm for solution of a two step problem of stochastic programming, *Cybernetics* 4 (1968) 48-50 MR 45.8268.

[S25] ———, ———, Some problems of approximation by functions with restrictions on the first and second derivatives, in, *Theory of Optimal Decisions*, Cybernetic Institute, (Kiev, 1976) (Russian).

[S26] ———, ———, An estimate of the rate of convergence of random search method, *Cybernetics* 10 (1974) 615-618 MR 52.16647.

[S27] ———, Convergence of a gradient type method with dilatation of the space in the direction of the difference of two successive gradients, *Cybernetics* 11 (1975) 564-570 MR 56.3719.

[S28] ———, L.P. Shabashova, The solution of minimax problems by the method of the generalized gradient descent with stretching of the space, *Cybernetics* 8 (1972) 88-94, MR 46.6595.

[S29] ———, L.A. Galustova, and A.I. Momot, Application of mathematical methods for optimal design of a single gas supply system considering dynamics of its development, *Cybernetics* 14 (1978) 68-73.

[S30] ———, Mathematical methods of investigation and optimization of economic systems, (Cybernetics Institute, Kiev, 1980) (Russian) MR 82h: 90002.

[S31] ———, I.M. Asel'derora, An investigation of an elementary pursuit problem on a network, *Cybernetics* 14 (1978) 550-555 MR 58.26066.

[S32] ———, *Minimization Methods for Nondifferentiable Functions and their Applications* (Naukova Dumka, Kiev, 1979) English translation to appear (Springer) MR 80f.49024. ISBN 0-387-12763-1 162pp. (1985)

[S33] ———, Application of generalized gradient descent in block programming, Cybernetics 3 (1967) 53-55.

[S34] ———, V.I. Gershovich, Solution of problems of high dimensionality by modified gradient minimization with space expansion, Cybernetics 17 (1981) 642-646.

[S35] S. Smale, The problem of the average speed of the simplex method, in [B3] 530-539.

[S36] V.A. Skokov, Note on minimization methods employing space stretching, Cybernetics 10, (1974), 689-692.

[S37] R.V. Southwell, Relaxation Methods in Engineering Science, (Oxford, 1940).

[S38] J. Spronk and J. Telgen, An ellipsoidal interactive multiple goal programming method, in J.Mark (ed.), Fourth International Conference on Multiple Criteria Decision Making (Springer, Berlin, 1981).

[S39] G.W. Stewart, Introduction to Matrix Computations (A.P., N.Y., 1973).

[S40] J. Stoer and C. Witzgall, Convexity and Optimization in Finite Dimensions I. Springer-Verlag, N.Y., 970

[S41] R.E. Stone, Khachiyan's algorithm with finite precision, Department of Operations Research, Stanford University, 1980.

[T1] S.P. Tarasov, and L.G. Hachiyan, Bounds of solutions and algebraic complexity of systems of convex diaphontine inequalities. SMD 22 (1980), 700-704.

[T2] R.E. Tarjan, Complexity of combinatorial algorithms, SIAM Review 20 (1978), 457-491.

[T3] J. Telgen, On relaxation methods for systems of linear inequalities, European Jr. of O.R. 9 (1982), 184-189.

[T4] ———, Redundancy and Linear Programming, Mathematical Center tracks no.137, (Amsterdam, 1981).

[T5] M.J. Todd, Some remarks on the relaxation method for linear inequalities, Technical Report 419, School of Operations Research and Industrial Engineering, Cornell University, Ithaca, New York (1979).

[T6] ———, On minimum volume ellipsoid containing part of a given ellipsoid, M.O.R 7 (1982), 253-261.

[T7] J.F. Traub and H. Wózniakowski, Complexity linear programming, O.R. Letters 1 (1982) 59-62.

[T8] K. Truemper, Algebraic characterization of unimodular matrices, SIAM Appl.Math. 35 (1978), 328-332.

[T9] ———, How do detect hidden networks and totally-unimodular subsections of linear programs? Technical Report, University of Texas at Dallas, 1983.

[T10] A.D. Tuniev, A generalization of the simplex method. Dokl. Akad. Nauk Arm. SSR 71 (1980) 198-203. (Russian) MR 83h. 90061, zbl 495.90054.

[T11] A.W. Tucker, Combinatorial theory underlying linear programs, Recent Advances in Mathematical Programming, R.E. Graves and P. Wolfe eds., (McGraw-Hill, N.Y., 1969) 1-16.

[U1] S. Ursic, The ellipsoid algorithm for linear inequalities in exact arishmetic. IEEE/FOCS (1982), 321-326.

[V1] C. van de Panne and F. Rahnama, The first algorithm for linear programming: an analysis of Kantorovich's method. Presented at the 10th Intern.Symp. on Mathematical Programming, Montréal, 1979; also Univ. of Calgary, Dept. of Economics, Discussion paper no.45,(1977).

[V2] A.F. Veinott and G.B. Dantzig, Integral extreme points, SIAM Review 10 (1968), 371-372.

[V3] N.O. Vil'chevskii, Choosing the penalty coefficient in linear programming problems. Automation and Remote Control 6 (1970) 626-632.

[V4] E.I. Volynskiy, I.A. Zil'berman, On the choice of penalty coefficients in iterative methods of linear programming. Engineering Cybernetics 16, no.3 (1978) 27-33.

[V5] J. von Neumann, A numerical method to determine optimum strategy, Naval Res.Log.Quart. 1 (1954), 109-115.

[V6] R. Voutilainen, The ellipsoid algorithm for linear programming. Arkhimedes 34, (1982) 88-96 (Finnish) MR 83k: 90063.

[W1] S. Walukiewicz, Ellipsoidal algorithm for linear programming, in [B3] 547-577.

[W2] D.S. Wang, Hachiyan's algorithm in VAXIMA, improvements and difficulties, in Lecture Notes in Computer Science no.144 (Springer, N.Y., 1982).

[W3] H.H. Weber, Khachiyan's algorithmus, Zeitschrift für O.R. B 26 (1982) 229-240.

[W4] K. Wendler, Die ε-störung linearer und quadratischer optimierungsaufgaben und ihre anwendung auf das Hildreth-Verfahren, Unternehmensforschung, 15 (1971), pp.1-14.

[W5] P. Wolfe , The composite simplex algorithm, SIAM Review 7 (1965) 42-54.

[W6] ———, A method of conjugate subgradient for minimizing nondifferentiable functions, MP Study 3 (1975), 145-173.

[W7] ———, A bibliography for the ellipsoid algorithm, IBM Research Center, Yorktown Heights, N.Y., USA.

[W8] ———, The ellipsoid algorithm, Science 208 (1980) 240-242.

[W9] ———, (Ed.), M.P. Study 3 (1975).

[Y1] M.Z. Yakın: On the convergence rate of a space extention algorithm for nondifferentiable optimization. University Houston, CBA WP 69, (1981).

[Y2] B. Yamnitsky, Notes on linear programming, Master's Thesis, Boston University (1982).

[Y3] ———, L.A. Levin, "A 1965 linear programming algorithm runs in polynomial time", in Proc. IEEE/FOCS, Chicago, 1982, also, in Waterloo Silver Jubilee Conf. on Combinatorics 1982.

[Z1] N. Zadeh, A bad network problem for the simplex method and other minimum cost flow algorithms, MP 5 (1973), 255-266.

[Z2] ———, What is the worst case behaviour of the simplex? Department of Operations Research, Stanford University, TR 27, 1980.

[Z3] I. Zang, M. Avriel, On functions whose local minima are global, JOTA 16, (1975), 183-190.

[Z4] ———, E.U. Choo, and M. Avriel, A note on functions whose local minima are global, JOTA 18, (1976), 555-559.

[Z5] ———, ———, ———, On functions whose stationary points are global minima, JOTA 22 (1977); 195-208.

[Z6] N.G. Zhurbenko, E.G. Pinaev, N.Z. Shor, G.W. Yun, Choice of fleet composition and allocation of aircraft to civil airline routes. Cybernetics 12 (1976) 636-640, also p.811.

[Z7] K. Zorychta, Improvements of the ellipsoidal algorithm for linear inequalities. Bull.Acad.Pol.Sci., Ser.Sci. Math. 30, 97-103 (1982). zbl 482.90059.

[Z8] S.I. Zuhovickii, and M.E. Primak, On the convergence of the method of Chevyshev centers and the method of centered sections for solving a convex programming problem. SMD 16 (1975) 615-618.

[Z9] J. Zur-Gathen, M. Sieveking, A bound on solutions of linear integer equalities and inequalities. Proceedings of AMS 72 (1978) 155-158.

[Z10] F.J. Gould, A homotopy algorithm for LP, presented at ORSA/TIMS meeting in Orlando, 1983.

[Z11] E.A. Gunn, An unconstrained dual for linear programming and its solution by piecewise linear optimization. Utilitas Mathematica 20 (1981), 5-19.

[Z12] C.L. Lemke, The constraint gradient method of linear programming. SIAM 9 (1961), 1-17.

[Z13] D.L. Mangasarian, Normal solutions of large scale linear systems. Technical Report, University of Wisconsin, Madison, 1983.

[Z14] J.F. Shapiro, Mathematical Programming: Structures and Algorithms, (Wiley, N.Y., 1979).

[Z15] ——— , Nondifferentiable optimization and large scale linear programming. Lecture Notes in Control and Information Sc. 14 (Springer, Berlin, 1979) MR 81.90131.

[Z16] J. Edmonds, Submodular functions, matroids and certain polyhedra, in Combinatorial Structures and their Applications (Gordon and Breach, N.Y., 1970).

[Z17] ———, R.G. Giles, A min-max relation for submodular functions on graphs, Annals of Discrete Mathematics 1 (1977) 185-204.

[Z18] D.J.A Welsh, Matroid Theory, (A.P., N.Y. 1976).

Index